Hummingbirds and Their Flowers

Hummingbirds
AND THEIR FLOWERS

KAREN A. GRANT AND VERNE GRANT

1968

COLUMBIA UNIVERSITY PRESS

New York and London

KAREN A. GRANT and VERNE GRANT collaborate in research on flower pollination. Verne Grant is now Professor of Biological Sciences, and Director of the Boyce Thompson Southwestern Arboretum, University of Arizona. He was formerly Professor in the Institute of Life Science at Texas A & M University; and Geneticist at Rancho Santa Ana Botanic Garden, California.

Research for this book was supported by the National Science Foundation through Grant GB-3620.

Library of Congress Catalog Card Number: 68-23462

Printed in Holland

Preface

This book-monograph was written in the hope that it would fill a gap in the literature on pollination of flowers by animals.

Among broad studies in this field, we have some excellent treatments of the various modes of pollination in particular geographical regions, from Müller's *Alpenblumen* (1881) to Vogel's floral ecological survey of South Africa (1954). The modes of pollination in particular plant groups have been treated in several monographs, from Darwin's *Fertilisation of Orchids* (1862) to our own *Flower Pollination in the Phlox Family* (1965). Insect behavior in relation to flower pollination has been dealt with by Frisch (1950, etc.), Knoll (1921-1926), Kugler (1955, etc.), and Kullenberg (1961).

But the treatment of particular pollination systems on a monographic scale is in an anomalous position. Bee pollination is the most thoroughly studied pollination system. However, the literature on this system consists, on the one hand, of scattered technical papers and, on the other, of excellent but relatively brief chapters in the more general works, such as those of Knuth (1906-1909), Kugler (1955), Percival (1965), Faegri and van der Pijl (1966), and others. Bird pollination was reviewed in a systematic and coherent manner by Porsch in his *Vogelblumenstudien* (1924-1929, 1926-1933). But this pioneering study could not of necessity be placed on as solid a foundation of observational data as might be desired.

We have, therefore, no broad comparative treatment of any single pollination system which is based on adequate field observations. The present monograph is an attempt to provide such a study.

PREFACE

The presentation of the subject matter in this book, in both text and color plates, has been designed so as to reach as wide a range of readers as possible—not only specialists in the field of pollination studies, but also general biologists and amateur naturalists.

The work described here was carried out while Verne Grant was a staff member of the Rancho Santa Ana Botanic Garden, Claremont, California. The research for this book was made possible by the financial support of the National Science Foundation and by the helpful interest of its administrative officials.

On three occasions we enjoyed the use of the facilities at the Southwestern Research Station of the American Museum of Natural History in southeastern Arizona. Much of the material in Chapter 9 was published earlier as a journal article in *Evolution*. Mr. Jesse Brock of Riverside Color Press, Riverside, California, was very helpful in the preliminary planning and organization of the color plates.

We wish to express our sincere gratitude to all the institutions, foundations, and individuals who have contributed in one way or another to the progress of this study.

KAREN A. GRANT

VERNE GRANT

vi

Contents

Hummingbirds and Their Flowers

1 *Introduction*

In many regions of the American hemisphere there exists a class of flowers adapted for pollination by hummingbirds. These flowers possess various characteristics which serve to attract hummingbirds, provide them with deep-seated nectar, and deposit pollen on parts of their bodies where it can be transferred to the stigma of another flower of the same species. The nectar-seeking birds bring about flower pollination in such species, and pollination in turn is a critical stage in the process of plant reproduction.

The hummingbirds for their part are specialized for feeding on flowers. With their long bills, extensile tongues, and ability to hover on the wing, these birds have successfully invaded the ecological niche of flower-visiting insects.

We have then, as between hummingbirds and their flowers, a relationship of co-adaptation involving partners which have attained a high level of specialization.

One of the questions which has to be asked about this mutual relationship between hummingbirds and hummingbird flowers is: how does it function? What role do hummingbirds play in pollination?

A second question is: by what pathways have the specialized floral mechanisms of hummingbird flowers developed in the course of evolution? Assuming that hummingbirds are an environmental factor in the lives of many plants, and that they exert a selective pressure on the flora, in what ways have some plants responded to this selection and become adapted to hummingbirds?

This question deals, however, with only one side of a co-adapted system. The hummingbirds have also gone through many generations of selection for and adaptation to flower feeding. And therefore a third problem is to explain the evolutionary development of the mutually adapted bird-flower relationship as a whole.

These questions have not been the subject of any broad and detailed study heretofore.

THE AREA

In attempting to deal with the questions posed above, it was necessary in the first place to block out for special study some one geographical segment of the whole area occupied by hummingbirds and their flowers. Only by restricting our fact-finding efforts in this way could we really come to grips with the problem and hope to bring the evolutionary picture into focus.

The area embraced in this study has different limits for different purposes. We have been able to carry out prolonged and detailed field studies in southern California, adequate studies in northern California and Arizona, and have gained some familiarity with the situation in several other western states.

On the basis of what we ourselves have learned at first hand, what is recorded in the literature, and what can be gleaned from museum collections, it proves possible for some purposes to generalize over the larger area of western North America. This is defined here as the region of the United States extending from the Pacific coast to the eastern base of the Rocky Mountains and from the Mexican to the Canadian borders.

British Columbia and Alaska have been brought into the story in a few instances, to show the continuation into the Pacific Northwest of geographical trends which are apparent within the American West.

OBJECTIVES

The first objective of this monograph is to present the known facts regarding hummingbird pollination in western North America. The types of

2

floral mechanisms, the systematic relationships of the western plants bearing hummingbird flowers, and their geographical and ecological distribution all come in for consideration (Chapters 3 to 7).

Our second purpose is to consider the evolution of hummingbird flowers in the American West (Chapters 7 to 10). In developing this theme we will treat the complex interrelationships between western hummingbirds and their flowers at three levels of biological organization.

At the level of individual plant species, various floral adaptations for hummingbird pollination have evolved independently in different plant genera and families. Next, at the community level, these bird-flowered plant species have influenced the character of plant communities from desert to timberline in the western United States. Finally, in the larger sphere of the ecosystem, certain characteristics shared in common by numerous unrelated bird-pollinated plant species have apparently evolved in response to the migratory habits common to several different hummingbird species.

Our third objective is a logical extension of the latter question. It is to consider in general terms the stepwise development to a high degree of specialization of a co-adapted system of flowers and pollinating animals. We wish to explain how plants can be responding to a selective factor in their biotic environment which is concurrently evolving in response to the same plants. In Chapter 11 we make a first attempt to deal with the poorly understood problem of reciprocal selection between plants and animals.

METHODS

The factual evidence is based primarily on field observations in natural populations. We have made numerous field trips to various parts of the West, and have observed the feeding behavior of hummingbirds on flowers at different times of day and in different seasons. In a few cases it has been possible to virtually live with the birds and their flowers for periods ranging from several days to a week or more. Such close and prolonged contact leads to an understanding of hummingbird pollination which can come in no other way. It has also been helpful to look for hummingbirds in nature at odd times of the year and in places where they are not expected to occur.

The field observations have been supplemented by more detailed examination of various aspects of the pollination process. Floral mechanisms have been worked out under the dissecting microscope. The mechanical insertion of the bill or head of a bird specimen into fresh flowers has often provided answers to questions about the pollination mechanism. Photographs of feeding birds taken through a telephoto lens have revealed and recorded characteristic behavioral traits of the birds which are relevant to their role as pollinators.

The photographs were taken with the object of showing the humming-birds and hummingbird flowers in their natural habitats. As we wished to illustrate the behavior of the birds as they fed on their flowers in nature, no special setups were ordinarily used. The rather simple photographic equipment (Exakta camera, 200 mm telephoto lens, and 1.85× tele-extender) was supplemented by many hours spent in the field. The procedure of sitting among the flowers for several hours on several consecutive days usually accustomed the hummingbirds to the nearby presence of the photographer. This, of course, was not always possible.

Behavioral differences between hummingbird species and between individuals of one species were noted in taking the pictures. For example, the Costa hummingbird proved the easiest to photograph, as it was not extremely wary and its natural habitat is in open country. On the other hand, the Calliope hummingbird was a very difficult subject, as it is both quick and shy and typically feeds and perches in deep shaded areas. Immature birds were less wary of being photographed than were the adults, and would often feed on flower after flower within a few feet of the camera.

This work has been underway for many years, during which it has gone through several phases. Preliminary field studies were made by the second author from 1949 to 1960. In 1960 the first author took up the problem and carried it along until 1964. Since then the study of hummingbird pollination in western North America has been the main project of the authors working jointly.

2 *The Hummingbirds*

GEOGRAPHICAL DISTRIBUTION

The hummingbird family, the Trochilidae, consists of 319 species grouped into 121 genera. This family of birds is restricted in distribution to the New World, where it ranges from Tierra del Fuego to Alaska and reaches its greatest abundance in the northern Andes. It is believed that the Trochilidae originated in South America and underwent a secondary radiation in North America (Mayr, 1964).

Hummingbirds are basically a tropical and subtropical group of birds, and gradually diminish in numbers of species at higher latitudes. These trends are shown in a generalized way in Map 1.

They are also most abundant in mountainous country. At tropical and subtropical latitudes in South America, hummingbirds exhibit an abrupt decrease in species density on passing from the Andes to the extensive eastern lowlands of that continent (Ridgway, 1891). A similar trend is seen at temperate latitudes in North America (Map 1). Seven species of hummingbirds belonging to four genera have breeding ranges largely north of the Mexican border in the mountainous western United States, whereas a single species, the Ruby-throated hummingbird (*Archilochus colubris*), occurs in the extensive lowland area of the eastern United States (Ridgway, 1891).

Most species of hummingbirds are resident, spending the entire year in a single region. But the species living in the southern and northern margins of the distribution area of the family have acquired the migratory habit.

5

MAP I.   NUMBER OF SPECIES OF HUMMINGBIRDS IN DIFFERENT LATITUDINAL ZONES

[Compiled and redrawn from Greenewalt (1960) and Austin (1961).]

GENERAL CHARACTERISTICS

Among birds, hummingbirds are unique in the structure of their wings (the primary quills are strongly developed while the secondary quills are much abbreviated) and possess long slender bills and extensile tongues of an unusual structure (Ridgway, 1891). Certain other anatomical features are common to swifts and hummingbirds and are developed to an extreme degree in both groups. Among these are the high keel to the sternum, the excessive development of the pectoral muscles, the short humerus and very long manus, and the small feet with relatively large curved sharp claws (Ridgway, 1891).

These and other skeletal characters reveal the close relationship between swifts and hummingbirds, which undoubtedly share a common ancestor. The present distribution of hummingbirds suggests that this family arose from a primitive swift-like progenitor in the American tropics (Austin, 1961).

The swifts are exceptionally fast fliers and live entirely on insects which they catch while in flight (Austin, 1961). Most hummingbird species feed on both insects and nectar as adults. The young nestlings are apparently given a diet consisting entirely of insects, and Wagner (1946) considers this together with the ontogeny of the bill, tongue, and stomach anatomy as showing the original food of hummingbirds to have been animal.

The reader is referred to the highly informative and interesting account by Ridgway (1891) for a discussion of the characteristics and habits of hummingbirds. The mechanism of hummingbird flight has been analyzed in detail by Greenewalt (1960) and some aspects of its physiology have been treated by Lasiewski et al. (1965). The highly modified tongue apparatus of hummingbirds has been described by Weymouth et al. (1964).

The Trochilidae exhibit a spectacular variety of often brilliantly colored irridescent gorgets, crests, or other ornamentations, and the great diversity among the group has been beautifully portrayed by Greenewalt (1960).

In size, hummingbirds range from the smallest known birds, the Bee hummingbird (*Calypte helenae*) of the West Indies which measures $2\frac{1}{4}$ inches from bill tip to tail tip, to the Giant hummingbird (*Patagona gigas*) of the Andes, with an $8\frac{1}{2}$-inch span from bill to tail (Ridgway, 1891).

7

A similar range of variation exists in bill size and shape. Hummingbird bills are slender and pointed, often straight, sometimes slightly curved or even sickle-shaped. Bill lengths vary from extremely short (6 mm in the Andean species, *Rhamphomicron microrhynchum*) to extremely long, as in the Sword-billed hummingbird (*Ensifera ensifera*) of Ecuador which has a bill more than 12.5 cm long (Ridgway, 1891).

The Trochilidae are polygamous and the female builds the nest, lays two eggs, incubates and raises the young nestlings. Territorial behavior is strongly developed in the group; the males establish posts from which they defend their territories from intruders. Various aspects of the life histories and behavior of hummingbirds have been treated in detail by many workers (see, for example, Bent, 1940; Bené, 1947, for North American species; Wagner, 1945, 1946, for Mexican species; and Skutch, 1931, 1951, 1958, 1964, 1967, etc., for Central American species).

FOOD REQUIREMENTS AND FEEDING HABITS

Hummingbirds feed at frequent intervals throughout the day, visiting flowers from just before dawn to just after sunset. At night, hummingbirds may become torpid, and through lowering of body temperature and metabolic rate conserve energy at a time when they are unable to feed for a long period.

The energy requirements of these small-bodied birds have been experimentally investigated by a number of workers (see inter alia, Pearson, 1950, 1954; Lasiewski, 1962, 1963). Pearson (1954) determined the time spent by a male Anna hummingbird (*Calypte anna*) at various activities in the relatively inactive post-breeding season and calculated the necessary energy requirements:

A male Anna hummingbird watched on September 3 and 8 flew on the average 18.7 per cent of the time. His energy exchange for 24 hours of normal life in the wild was calculated to be 7.55 Calories (assuming torpidity at night) or 10.32 Calories (assuming sleep at night). During his 12-hour 52-minute active day, most of his energy expenditure was distributed as follows: perching, 3.81 Calories (56 per cent); nectar flights, 2.46 (36 per cent); insect-catching flights, .09 (1.3 per cent); and defense of territory, .30 (4.5 per cent). The nectar secretion of about 1022 Fuchsia blossoms can supply this daily need.

The diet of hummingbirds consists of floral nectar and small insects,

8

spiders, or other animal life obtained from flowers, on the bark or leaves of trees, or hawked in the air. It has long been debated whether insects or nectar form the principal food of these birds; most species apparently utilize both food types, although the extent to which they rely on one or the other may differ between species.

In Mexico, Wagner (1946) found two species, the Cinnamomeous hummingbird (*Amazilia rutila*) and Prevost's Mangos (*Anthracothorax prevostii*), which were predominantly nectarivorous, while Abeille's hummingbird (*Abeillia abeillei*) apparently depended mostly on insect food. Wagner (1946) concluded that the diet of most of the Mexican species which he observed included both food types, although a given species might feed mainly on one or the other in different seasons, according to availability.

The western North American hummingbirds similarly feed on both insects and nectar, although the Anna hummingbird apparently relies more heavily on insect food than do the other species (see Pitelka, 1942), and this habit may explain why *Calypte anna* is able to remain in California throughout the winter season when floral food is mostly unavailable.

HUMMINGBIRD TONGUE

Recent investigations on the structure of the Trochilid tongue reveal that it is not a hollow tube-like "sucking" organ as has been generally believed. The highly modified hyoid apparatus and its musculature in hummingbirds, a group with highly specialized feeding habits, is described by Weymouth et al. (1964).

Morphological studies of the tongue of *Selasphorus sasin* confirmed the earlier but largely overlooked work of Scharnke that the two parallel internal chambers of the tongue are not hollow tubes and do not even open to the outside as was earlier believed (Weymouth et al., 1964). The hummingbird tongue forks distally and each half forms a membranous curled trough. This membranous layer is sometimes fimbriated, probably due to wear and tear. Capillary action apparently carries the nectar into the external troughs of the tongue, and when the tongue is retracted into the mouth the nectar is swallowed in the usual way. Small insects may become entangled in the fimbriated tongue tip as the birds probe in flowers for nectar.

9

THE WESTERN NORTH AMERICAN SPECIES

Seven species of hummingbirds have breeding ranges largely within western North America. These are the Broad-tailed, Rufous, Allen, Calliope, Black-chinned, Costa, and Anna hummingbirds (Table 1). In this book we are concerned mainly with the seven species just named, which will be described in more detail below.

TABLE 1. HUMMINGBIRD SPECIES OCCURRING IN WESTERN NORTH AMERICA

I. Species with breeding ranges lying largely in the western United States and Canada

Selasphorus platycercus	Broad-tailed hummingbird	Plate 16D
Selasphorus rufus	Rufous hummingbird	Plate 15A–C
Selasphorus sasin	Allen hummingbird	Plate 13
Stellula calliope	Calliope hummingbird	Plate 15D
Archilochus alexandri	Black-chinned hummingbird	Plate 14
Calypte costae	Costa hummingbird	Plate 11
Calypte anna	Anna hummingbird	Plate 12

II. Mexican species whose breeding ranges extend into the southern margins of the western United States

Lampornis clemenciae	Blue-throated hummingbird	Plate 16A
Eugenes fulgens	Rivoli hummingbird	Plate 16B
Cynanthus latirostris	Broad-billed hummingbird	Plate 16C
Amazilia violiceps	Violet-crowned hummingbird	

III. Mexican species which occur as rare visitors in the southwestern United States

Calothorax lucifer	Lucifer hummingbird
Hylocharis leucotis	White-eared hummingbird

In addition, several Mexican species extend their breeding ranges into the southern margins of Arizona, New Mexico, and Texas. Among these are the Blue-throated, Rivoli, Broad-billed, and Violet-crowned hummingbirds (Table 1). These species enter our story too, but not to the same extent as the seven truly western American species.

Feeding on small insects and floral nectar, the seven species of western American hummingbirds mostly migrate northward to their breeding ranges in early spring as the flowering season commences, and return to their wintering ranges southward in late summer and early fall. They breed in habitats ranging from the arid and desert country of the Southwest to the subalpine zone in the higher mountains. Some of them have breeding ranges extending over much of the western United States, while the breeding ranges of several are narrowly circumscribed.

In contrast to the diversity displayed among tropical hummingbirds, the western North American species are uniformly small in size and all display gorgets which are red to violet in color. Their bills are straight or very slightly curved and about 17 to 21 mm long in most cases; the smallest species, the Calliope hummingbird, has a slightly shorter bill which measures 15 to 17 mm. (These figures are based on measurements of about 20 specimens of each western species except the Broad-tail in the Los Angeles County Museum.)

The genus Selasphorus is represented in the western United States by three species. The Broad-tailed hummingbird (*Selasphorus platycercus*) is the characteristic hummingbird of the Rocky Mountain region. The breeding range of this summer-resident mountain inhabitant extends west to east-central California and south to Arizona, Texas, and Mexico. In the winter season Broad-tailed hummingbirds are concentrated in west-central Mexico (see Bent, 1940, for more detailed data of breeding and wintering ranges of the North American hummingbird species).

The Rufous hummingbird (*Selasphorus rufus*), like the Broad-tail, has a markedly disjunct winter and breeding range. This species migrates northward from south and central Mexico in the spring, following a route along the Pacific coast to its breeding range in the Pacific Northwest. Nesting in the region extending from southern Montana, Idaho, and Oregon to southeastern Alaska, the Rufous hummingbird claims the northernmost limit of distribution among the Trochilidae. In the late summer, the chief route of the south bound birds is along the summit of the Rocky Mountains (Phillips, Marshall, and Monson, 1964).

The Allen hummingbird (*Selasphorus sasin*) has a narrow breeding range along the California coast from Ventura County to Oregon, wholly within the humid coastal fog belt. A non-migratory race (*Selasphorus sasin sedentarius*) is resident on the offshore islands of southern California. The mainland race winters in Mexico.

The Calliope hummingbird (*Stellula calliope*) is the smallest of the North American hummingbirds. It is found nesting up to timberline in the higher mountains of western North America. A spring migrant from southern Mexico, the Calliope breeds from northern Baja California north to southern British Columbia and east to Montana, Wyoming, and Utah.

The breeding range of the Black-chinned hummingbird (*Archilochus alexandri*) covers an extensive portion of western North America, extending from northwestern Mexico and northeastern Baja California north along the Pacific coast to southwestern British Columbia and east to western Colorado, New Mexico, and western Texas. This species, however, is abundant only in the southern part of its breeding range (Bent, 1940), where it commonly nests in dry foothill and canyon regions at lower elevations. The winter range of the Black-chinned hummingbird extends from extreme southern California to southern Mexico.

The Costa hummingbird (*Calypte costae*) is a species of desert and arid foothill regions, breeding from southern Baja California north to southern California and southern Nevada, and east to southwestern Utah, east-central Arizona, and southwestern New Mexico. This species is slightly migratory, wintering from southern California and southern Arizona south to southern Baja California.

The Anna hummingbird (*Calypte anna*) breeds from northern Baja California north nearly to the Oregon border and entirely west of the Sierra Nevada crest. It is resident in California, withdrawing only from the extreme northern part of its breeding range in winter. Some individuals of this species winter in southern Arizona, apparently returning to California prior to the late winter start of the breeding season (Phillips, Marshall, and Monson, 1964).

The breeding ranges of most of the hummingbird species of western America overlap in various combinations, so that one species occurs with one or more other species. In regions where two or more breeding species occur together, differences in habitat preference promote separation of the species. However, overlaps in habitat distribution also occur (Pitelka, 1951b).

Pitelka (1951b) found four hummingbird species (the Anna, Allen, Black-chin, and Costa) breeding near Santa Barbara, California. An overlap in habitat distribution of nesting females occurred in the Anna, Allen, and Black-chin. The breeding seasons of these four species were found to overlap from mid-April through June. The Anna hummingbird normally breeds from December into June, raising two, or perhaps three broods. The Allen raises two broods (February to June) while the Black-chin and the Costa each raise a single brood from late April through June.

In California, the Calliope hummingbird raises a single brood in June and July (Dawson, 1923); the Rufous hummingbird apparently breeds in late spring (Bent, 1940), while the Broad-tail raises a single midsummer brood (Bené, 1947).

The plantings in the Rancho Santa Ana Botanic Garden are grouped according to the main plant communities of California. Two species of hummingbirds are common in this 85-acre garden. It is significant that each species of hummingbird occurs in greatest numbers in the artificial communities which correspond to its natural habitat. Thus the Costa hummingbird is found mainly in the open brushy parts of the Botanic Garden, and the Anna hummingbird in live oak woods. The Rufous hummingbird, which passes through in the early spring, is usually seen feeding on flowers growing in the deep shade of tall trees. This microdistribution probably reflects the different ecological preferences of the different hummingbird species.

GEOGRAPHICAL TRENDS IN ABUNDANCE IN WESTERN NORTH AMERICA

Let us compare different regions of the West and Northwest with respect to the number of breeding species of hummingbirds. Using the distributional data given by Bent (1940) and in some cases by later authors, we have listed the numbers of breeding species in the various western and northwestern states and provinces. The results are summarized in Table 2 and shown graphically in Map 2.

It is seen that the number of hummingbird species having breeding ranges within a state decreases from seven or six in the southwestern United States to three in the Pacific Northwest and one in Alaska. The number also

13

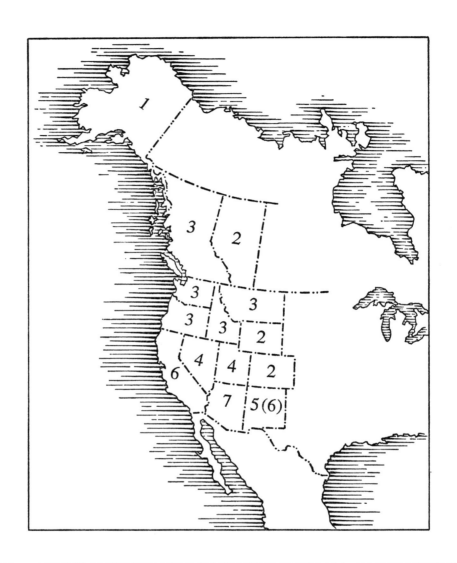

MAP 2. NUMBER OF HUMMINGBIRD SPECIES WITH BREEDING RANGES IN DIFFERENT
WESTERN AND NORTHWESTERN STATES

TABLE 2. THE NUMBER OF HUMMINGBIRD SPECIES BREEDING IN DIFFERENT REGIONS OF WESTERN AND NORTHWESTERN NORTH AMERICA

(Compiled mainly from data of Bent, 1940)

Region	Number of Breeding Species	Bird Species
Alaska	1	Rufous
Alberta	2	Rufous, Calliope
Arizona	7	Costa, Black-chin, Broad-tail, Rivoli, Broad-bill, Blue-throat, Violet-crown
California	6	Anna, Allen, Costa, Black-chin, Calliope, Broad-tail
British Columbia	3	Rufous, Calliope, Black-chin
Colorado	2	Broad-tail, Black-chin
Idaho	3	Rufous, Calliope, Black-chin
Montana	3	Rufous, Calliope, Black-chin
Nevada	4	Broad-tail, Calliope, Black-chin, Costa
New Mexico	5 (6)	Broad-tail, Black-chin, Rivoli, Blue-throat, Costa, Broad-bill (?)
Oregon	3	Rufous, Calliope, Black-chin
Utah	4	Broad-tail, Calliope, Black-chin, Costa
Washington	3	Rufous, Calliope, Black-chin
Wyoming	2	Calliope, Broad-tail

decreases toward the interior of the continent. These trends are obviously a continuation of those found in the family Trochilidae as a whole and mentioned earlier.

Furthermore, there is a difference between southern and northern areas in the abundance of hummingbird individuals which is greater than the figures in Table 2 indicate. There are two reasons for this.

In the first place, some species breed over a wide range, but are abundant only in the southern part of their breeding range. This is the case with the Black-chinned hummingbird, which has a breeding range extending over much of western North America but is abundant only in the Southwest. Similarly, the Costa, though extending into Utah, Nevada, and rarely into

New Mexico, is primarily a species of the California and Arizona deserts (Bent, 1940).

Second, the species with northern breeding ranges migrate through the southern areas on their way to and from their winter range in Mexico, thus adding to the total number of individual hummingbirds in the southern states. The cumulative effect of migrant species on the density of hummingbird populations can be illustrated by the situation in Arizona. Seven species of hummingbirds breed in Arizona. But four other species occur in that state, some as common migrants. The Rufous, Calliope, and Allen hummingbirds pass through Arizona in spring and fall, while the Anna hummingbird spends fall and winter there (Phillips, Marshall, and Monson, 1964).

3 *Hummingbird Flowers of Western North America*

In this chapter we present a systematic list of plant species which occur in western North America and bear hummingbird flowers. In drawing up this list, western North America was defined arbitrarily as the area of the United States between the Pacific coast and the Rocky Mountains and between the Mexican border and the Canadian border. By hummingbird flowers we mean flowers adapted primarily for feeding of and pollination by hummingbirds, as described in Chapter 4.

The list contains 129 plant species which fulfill the stated geographical and floral biological criteria. For 41 of the species we have field evidence of hummingbird pollination; these are annotated with an R in the list. The documentary details have been published elsewhere and will not be repeated here (see Grant and Grant 1966, 1967a, 1967b). The remaining 86 species have flowers of a similar type, suggesting strongly a similar mode of pollination. Many doubtful or borderline cases were considered but rejected in compiling the list. Therefore the list errs on the conservative side and there are probably more than 129 species of hummingbird flowers in the western American flora.

The taxonomic information in the list was obtained from various state and regional floras (Abrams, 1940-1960; Davis, 1952; Harrington, 1954; Hitchcock, Cronquist, Ownbey and Thompson, 1955-1964; Jepson, 1922-1943, 1925; Kearney and Peebles, 1960; Munz and Keck, 1959; Peck, 1961; Wooton and Standley, 1915). In matters of nomenclature and taxonomic concepts we have usually (but not invariably) followed the more recent authors.

The species are grouped by families, and the families are grouped in turn into one natural and two partly artificial assemblages. The three main categories recognized for purposes of convenience are the Monocotyledons, the Dicotyledons with free or no petals, and the sympetalous Dicotyledons with fused petals.

SYMPETALOUS DICOTYLEDONS

Fouquieriaceae

FOUQUIERIA

F. splendens. Ocotillo. Deserts from southeastern California to western Texas, and northern Mexico. (R)* Plates 1A, 8A.

Convolvulaceae

IPOMOEA (Morning-glory)

I. coccinea. Star-glory. Arizona to western Texas, south to tropics.

Polemoniaceae

COLLOMIA

C. rawsoniana. South-central Sierra Nevada, California.

GILIA

G. splendens. San Bernardino Mountain race. High San Bernardino Mountains, southern California. (R)

G. subnuda. Northern Arizona to Nevada, east to New Mexico and Colorado.

IPOMOPSIS

I. aggregata. Widespread in western North America. (R) Plate 8B.

I. arizonica. Eastern California to Arizona and Utah. (R)

I. tenuifolia. Extreme southern California and northern Baja California.

POLEMONIUM

P. pauciflorum. Southeastern Arizona and northern Mexico.

P. brandegei. Rocky Mountains, Colorado.

Labiatae

MONARDA (Bee-balm)

M. fistulosa. Rocky Mountain region from Canada to Arizona, and eastern United States. (R; see Loew, 1904-1905)

---

*R ≡ pollination records available.

MONARDELLA
M. macrantha. Central California to northern Baja California. (R) Plate 7C.

SATUREJA
S. mimuloides. Central and southern California.

SALVIA (Sage)
S. henryi. Southern Arizona to western Texas and Mexico.
S. lemmoni. Southern Arizona and northern Mexico. (R) Plate 7A.
S. spathacea. Pitcher-sage. California Coast Range. (R) Plate 7B.

STACHYS (Hedge-nettle)
S. chamissonis. Coastal California.
S. ciliata. Oregon to British Columbia.
S. coccinea. Southern Arizona to western Texas and Mexico.

TRICHOSTEMA
T. lanatum. Wooly Blue-curls, Romero. Coast Ranges of central and southern California. (R) Plate 7D.

Scrophulariaceae

CASTILLEJA (Indian Paint Brush)
C. affinis. California. (R; Merritt, 1897)
C. angustifolia. Eastern Oregon to southern British Columbia and northwestern Wyoming.
C. applegatei. Central California to eastern Oregon, east to western Wyoming. (R)
C. austromontana. Southern Arizona, southern New Mexico, and northern Mexico.
C. brevilobata. Northern California and southern Oregon.
C. breweri. Sierra Nevada, California. (R)
C. chromosa. Southern California to eastern Oregon, east to New Mexico, Colorado, and Wyoming. Plate 6D.
C. covilleana. Central Idaho and western Montana.
C. crista-galli. Idaho to western Wyoming.
C. cruenta. Southeastern Arizona.
C. culbertsonii. Sierra Nevada, California.
C. elmeri. Cascade Mountains, Washington.
C. exilis. Widespread in western North America.
C. foliolosa. California. (R)
C. franciscana. Central Coast Range, California.
C. fraterna. Wallowa Mountains, Oregon.
C. haydeni. Northern New Mexico to Colorado.

HUMMINGBIRD FLOWERS

C. hispida. Oregon to British Columbia, east to Montana.

C. hololeuca. Channel Islands, California.

C. inconstans. Northern New Mexico.

C. integra. Arizona to western Texas, south to northern Mexico, and north to Colorado.

C. lanata. Arizona and northern Mexico to western Texas.

C. latifolia. Monterey coast, California.

C. laxa. Southern Arizona and northern Mexico.

C. lemmonii. Sierra Nevada, California.

C. leschkeana. Point Reyes, California.

C. linariaefolia. California to Oregon, east to New Mexico, Colorado, and Wyoming. (R) Plate 6C.

C. martinii. Southern California to Baja California. (R)

C. miniata. Widespread in western North America. (R) Plate 6B.

C. minor. Arizona, New Mexico, and northern Mexico.

C. nana. Sierra Nevada, California.

C. neglecta. Tiburon peninsula, California.

C. organorum. Organ Mountains, New Mexico.

C. parviflora. Oregon to Alaska and northern Rocky Mountains.

C. patriotica. Southeastern Arizona and northern Mexico. (R)

C. payneae. Cascade Mountains from northern California to central Oregon. (R) Plate 6A.

C. peirsonii. Sierra Nevada, California.

C. plagiotoma. San Gabriel Mountains, California.

C. pruinosa. Central California to Oregon.

C. rhexifolia. Northern Oregon to British Columbia, east to Colorado and Alberta.

C. roseana. Coast Range, California.

C. rupicola. Oregon to British Columbia.

C. stenantha. California. (R)

C. suksdorfii. Oregon to British Columbia.

C. subinclusa. Sierra Nevada, California.

C. uliginosa. Pitkin Marsh, north-central California.

C. wightii. Coast line from central California to Washington.

C. wootoni. White and Sacramento Mountains, New Mexico.

DIPLACUS

D. aurantiacus. Bush Monkey-flower. California and Oregon. (R)

D. flemingii. Channel Islands, California.

D. puniceus. Coastal southern California and northern Baja California. (R)

GALVEZIA

G. speciosa. Islands of southern California and Baja California. (R) Plate 5E.

20

KECKIA (= Penstemon section Hesperothamnus, see Straw, 1966.)

K. cordifolia. Central California to northern Baja California. Plate 5F.

K. corymbosa. Northern California.

K. ternata. Southern California and northern Baja California. (R)

MIMULUS (Monkey-flower)

M. cardinalis. Crimson Monkey-flower. California to Oregon, east to Arizona and Nevada. (R) Plate 5D.

M. eastwoodiae. Northeastern Arizona and southeastern Utah.

PEDICULARIS (Lousewort)

P. densiflora. Indian Warrior. California to southern Oregon and northern Baja California. (R)

PENSTEMON (Beard-tongue)

P. barbatus. Arizona to Utah and southern Colorado, and south into Mexico. (R) Plate 5A.

P. bridgesii. California and Baja California to Colorado and New Mexico. (R)

P. cardinalis. South-central New Mexico.

P. centranthifolius. Scarlet Bugler. Coast Range, central California to Baja California. (R) Plate 5B.

P. clevelandii. Southern California and Baja California. (R)

P. crassulus. Central New Mexico.

P. eatonii. Southern California to Arizona and Utah.

P. labrosus. Southern California and northern Baja California. (R)

P. lanceolatus. Southeastern Arizona and southwestern New Mexico, to northern Mexico.

P. newberryi. Mountain Pride. Central California to northern Oregon. (R) Plate 5C.

P. parryi. Southern Arizona and northern Mexico.

P. pinifolius. Southeastern Arizona and southwestern New Mexico, to northern Mexico.

P. rupicola. Northern California to Washington.

P. subulatus. Arizona.

P. utahensis. Eastern California to northern Arizona and Utah.

SCROPHULARIA (Figwort)

S. coccinea (= *S. macrantha*). Southwestern New Mexico.

Acanthaceae

ANISACANTHUS

A. thurberi. Chuparosa, Desert Honeysuckle. Arizona, southwestern New Mexico, and northern Mexico. (R) Plate 8E.

21

HUMMINGBIRD FLOWERS

BELOPERONE

B. californica. Chuparosa. Deserts of southern California, Arizona, and northern Mexico. (R) Plate 8C–D.

JACOBINIA

J. ovata. Southern Arizona and northern Mexico.

Rubiaceae

BOUVARDIA

B. glaberrima. Southern Arizona and New Mexico to northern Mexico. (R) Plate 8F.

Caprifoliaceae

LONICERA (Honeysuckle)

L. arizonica. Arizona, New Mexico, and Utah.

L. ciliosa. Northern California to British Columbia, east to Montana.

L. involucrata ledebourii. California coast. (R) Plate 9A.

Campanulaceae

LOBELIA

L. cardinalis. Cardinal-flower, Scarlet Lobelia. Widespread in the southwest, eastern United States, Mexico, and Central America. (R)

L. laxiflora. Southern Arizona to Central America.

Onagraceae

ZAUSCHNERIA (California Fuchsia)

Z. californica. California to southern Oregon, east to New Mexico, south to northern Mexico. (R) Plate 9B.

Z. cana. Central and southern California.

Z. garrettii. Eastern California to Utah and western Wyoming.

Z. septentrionalis. Northern California.

CHORIPETALOUS AND APETALOUS DICOTYLEDONS

Nyctaginaceae

ALLIONIA

A. coccinea. Arizona, New Mexico, and northern Mexico.

Caryophyllaceae

SILENE (Catchfly)

S. californica. Indian Pink. Central California to southern Oregon.

S. laciniata. California to western Texas and Mexico. (R) Plate 10C.

Ranunculaceae

AQUILEGIA (Columbine)

A. desertorum. Northern Arizona.

A. elegantula. Arizona and New Mexico, north to southern Utah and Colorado, and south to northern Mexico.

A. eximia. Coast Range, California.

A. formosa. California to Alaska, east to Utah and Montana. (R) Plate 10A.

A. shockeyi. Desert mountains of eastern California and Nevada.

A. triternata. Eastern Arizona to western New Mexico and western Colorado.

DELPHINIUM (Larkspur)

D. cardinale. Scarlet Larkspur. Central California to Baja California. (R) Plate 10B.

D. nudicaule. Central California to southern Oregon.

Saxifragaceae

RIBES (Currant, Gooseberry)

R. speciosum. Fuchsia-flowered Gooseberry. Coast from central California to northern Baja California. (R) Plate 9C.

Papilionaceae

ASTRAGALUS (Locoweed, Rattleweed)

A. coccineus. Desert mountains of eastern California, southwestern Arizona, and northern Baja California. Plate 9D.

ERYTHRINA (Coral-tree)

E. flabelliformis. Coral-bean, Chilicote. Southern Arizona, southwestern New Mexico, and northern Mexico.

Cactaceae

ECHINOCEREUS (Hedgehog Cactus)

E. triglochidiatus. Southeastern California to New Mexico and Colorado, south to northern Mexico. (R) Plate 9E–F.

23

MONOCOTYLEDONS

Liliaceae

BRODIAEA
B. ida-maia. Fire-cracker Plant. Northern California to southern Oregon. Plate 10D.
B. venusta. Northwestern California.

FRITILLARIA
F. recurva. Scarlet Fritillary. Central California to southern Oregon.

LILIUM (Lily)
L. maritimum. Northern coastal California.
L. parvum. Alpine Lily. Sierra Nevada, California, to southern Oregon. (R)

Agavaceae

AGAVE (Maguey)
A. schottii (?). Southern Arizona to southwestern New Mexico and northern Mexico.
A. utahensis (?). Southeastern California to northern Arizona and southern Utah.

4 *Floral Mechanisms*

A particular association of floral features is commonly found in the western North American hummingbird flowers. These are usually solitary or loosely clustered flowers borne in a pendant or more or less horizontal position at the tips of flexible pedicels. The flowers are thick-tissued, often red or red combined with yellow, and yield large quantities of nectar at the base of a long stout floral tube. In considering these features we will follow the classification of Straw (1956b) with certain modifications.

Floral mechanisms can be grouped into four general classes according to their function in contributing to the efficiency of pollination. *Attractive mechanisms* are those floral features which serve to attract hummingbirds to the flowers. Characteristics of the flowers which inhibit flower visitors other than hummingbirds will be treated here as *exclusive mechanisms*. *Protective mechanisms* include features which serve to prevent injury to the flowers by the probing of the sharp-billed hummingbirds. *Pollinating mechanisms* involve the spatial and temporal relations of the reproductive organs so that feeding hummingbirds bring about pollination.

ATTRACTIVE MECHANISMS

Floral odors and floral colors generally serve to attract animal pollinators from a distance, while floral foods stimulate repeated visits to the flowers once the food source is discovered.

Hummingbirds are not perceptive of odors. Correspondingly, odors are frequently lacking in western American hummingbird flowers.

Hummingbirds do discriminate between colors. The colors of western

25

American bird flowers are predominantly some shade of red, or red combined with yellow. The red color is usually found in the corolla (Penstemon, Mimulus) (Plate 5). In Castilleja the flowers are not showy, but the leafy bracts subtending them are large and red (Plate 6). The stamens may also contribute bright colors, as in Aquilegia and Ribes (Plates 9, 10).

The occurrence of red as the predominant floral color is probably associated with the fact that, while hummingbirds perceive this color, bees do not. The problem will be discussed more fully in Chapter 10.

Hummingbirds also discriminate between the tastes of different concentrations of sugar syrups, and between syrups of honey or brown sugar or white sugar (see Bené, 1947).

Western American bird flowers are characteristically open throughout the day and yield large quantities of nectar. Their provision of an abundant nectar source which is available throughout the day and easily accessible to the hummingbirds is the primary factor attracting repeated visits of the birds.

The bills of most of the hummingbird species which breed in western North American range from 17 to 21 mm long from tip to feathered region, or are slightly shorter (15 to 17 mm) in the Calliope. The tongue can be extended varying distances beyond the tip of the bill. Throughout most of its length the bill is slender, about 1.5 mm in diameter in the Anna.

The nectar in western hummingbird flowers is furnished in tubes which correspond well in size and proportions with these bird bills. Representative measurements of the length and diameter of the floral tube or spur in various western species of bird flowers are given in Table 3. It will be seen that the floral tubes of most of the species listed range from 15 to 25 mm. long, and are somewhat stouter than a hummingbird bill.

These or any other bare measurements can tell only part of the story. Nectar accumulates to varying levels in the base of the floral tube, reducing the distance from orifice to nectar accordingly. Also the bird can stand off or probe deeply, and its tongue is capable of variable extension. There is consequently considerable free play between bill and floral tube.

The correspondence between hummingbird bills and floral tubes is best revealed by field observations of birds feeding on particular species of flowers. In a simple case exemplified by *Penstemon centranthifolius* we see bills 17 to 21 mm. long with extensile tongues probing in corolla tubes 22 to 25 mm long. Several different cases of mutual correspondence between feeding birds and flowers are recorded by photographs in Plates 18 to 24.

EXCLUSIVE MECHANISMS

Pollination efficiency is increased with the reduction of competition for nectar by non-pollinating flower visitors. The placement of the nectar at the base of long floral tubes in hummingbird flowers makes it mostly inaccessible to any but long-tongued insects. The elongated floral tubes described in the preceding section serve an exclusive as well as an attractive function. The chief competitors for the nectar of the day-blooming hummingbird flowers are long-tongued bees and butterflies.

Bees are attracted to flowers by their colors and odors. The frequent absence of floral fragrances and the occurrence of red floral colors are features of hummingbird flowers which lack attractiveness to bees.

The nectar of these flowers, accessible to hummingbirds which feed from a hovering position, is often made inaccessible to bees and butterflies, which must alight on flowers to feed. In flowers pollinated by bees or butterflies, landing platforms are often formed by the erect position and spreading petals of large radially symmetrical flowers, or the clustering of small flowers into heads, or by the conformation of the lower petals of bilaterally symmetrical flowers.

These alighting surfaces are eliminated in hummingbird flowers through various floral modifications. Among those species with a radially symmetrical floral organization the landing platform may be eliminated by the pendant position of the flowers (as in hummingbird-pollinated Aquilegia species) or by the recurving of the petals (*Ipomopsis aggregata*). Bilaterally symmetrical hummingbird flowers are often borne in a more or less horizontal position and may have the petals of the lower lip much shortened or strongly recurved (*Penstemon centranthifolius*, *Mimulus cardinalis*). A less effective mechanism occurs in *Trichostema lanatum*, where the long exserted stamens are strongly recurved, blocking the entrance to the corolla tube.

27

PROTECTIVE MECHANISMS

Flowers are commonly injured by the birds which feed on them, and the development of thick and strong-tissued floral parts is associated with bird pollination. Mechanical strengthening of flowers may be achieved by the presence of sclereids, or collenchymatous or sclerenchymatous tissues in various floral parts (see review by Grant, 1950).

The hummingbird-pollinated species of western North America typically bear thick-tissued flowers; their position at the tips of flexible pedicels aids in protecting them from injury by hummingbirds probing them with their bills.

The inner floral organs of flowers visited by birds are also subject to injury. If bird flowers are surveyed the world over, a high percentage of them reveal some special means of ovule protection. This protection is afforded by the separation of the nectary from the ovary and may be achieved by the ovary being elevated above the nectary on a stipe, or by a stamen column serving as a sheath between the ovary and nectary, or by an inferior position of the ovary (Grant, 1950).

The subsample of western North American hummingbird flowers does not show a similar high frequency of such floral structures serving to protect the ovules. Here the plant species are concentrated in the Scrophulariaceae and related families (see Chapter 8), which do not possess these particular means of ovule protection. The hummingbirds' bills are diverted away from the ovary and to the nectary in other ways among these species. Several examples may be given.

The two-lipped flowers of *Beloperone californica* (Acanthaceae) are about 3 cm long and tubular (Plate 8C-D). The two stamens and the style lie along the upper side of the corolla above the entrance to the corolla tube. The style is positioned at the middle of the upper side of the corolla in a groove formed between two ridges of corolla tissue; the two stamens depart at the base of the tube and are positioned in grooves along the side of the tube. The ridges on either side of the grooves containing the style and stamens join in a pair of rib-like projections on either side of the ovary, the latter being seated in a recess behind these projections. A fourth groove is situated on the inside lower surface of the tube, running from the corolla entrance to the nectar chamber. Thin sweet nectar is secreted by a cushion-like nectary surrounding the base of the ovary and spreads out in a film on the numerous small hairs which cover the inside of the tube and the rib-like projections just above the ovary.

It can be determined that the bird's bill is guided down the lower groove to the nectary without injuring the ovary by inserting a dissecting needle down the tube. The needle is guided by the lowermost groove and the rib-like projections into the nectar chamber at a level below the ovary. The probing object makes contact with the ovary as shown by first putting some sticky clay on the needle point, and after probing the tube, finding it adhering to the lower wall of the ovary. Repeated probes with a sharp-pointed needle left no prominent scratches on the ovary wall, showing that the ovary received only a glancing blow.

That the probing bills of hummingbirds similarly reach the nectar without striking the ovary with more than a harmless glancing blow is indicated by the fact that no evidence of damage to the ovaries was found in a score of flowers dissected and examined from a colony which had been extensively worked over by hummingbirds.

Thus, in *Beloperone californica* the female reproductive organ is protected by the position of the style within a groove in the corolla tube wall, the projecting ribs which shield the ovary from above, and the lower groove which guides the bird's bill to the nectar below the ovary.

A different mechanism occurs in the genus Penstemon where the site of the nectaries has shifted from the base of the ovary to the outer bases of the upper pair of stamens. This separation of the nectaries and the ovary together with the positioning of the broadened stamen bases between the ovary and the probing bill of a bird serve to protect the ovules of hummingbird-pollinated Penstemon species (Straw, 1956b).

Kerner (1894-1895) pointed out that pollen is often exposed in plants which bloom during a rainless season, but is protected from rain by various floral devices in plants that bloom in a rainy climate. We see examples of this rule in western American hummingbird flowers. The corolla forms a roof over the pollen in *Ipomopsis arizonica* and *Penstemon barbatus* which grow in regions of summer rain in the Southwest (Plate 5A). By comparison, related species of summer-dry areas in California either have exposed stamens, as in *Ipomopsis aggregata*, or lack a strongly developed corolla shield over the stamens, as in *Penstemon centranthifolius* (Plate 5B). Pollen protective mechanisms in hummingbird flowers warrant further study.

POLLINATION MECHANISMS

For pollen to be transferred the reproductive organs must be situated within the flower where they contact the body of the hummingbird as it probes for nectar. Among the typical hummingbird flowers of western North America, a spatial separation of the accumulated nectar and essential organs results in the deposition of pollen on the feathered parts of the bird's body where it adheres readily.

Various floral modifications accomplish this spatial separation. In hummingbird-pollinated species of Aquilegia and Delphinium (Ranunculaceae) nectar accumulates at the tips of long floral spurs.

The pendant red and yellow flowers of *Aquilegia formosa* possess five vertically held spurs and a central stamen column. Hummingbirds hover beneath these flowers and probe upward into the spurs; pollen is dusted on the backs of their heads or on their chins and throats, depending on whether they are oriented dorsally or ventrally with respect to the stamen column (Grant and Grant, 1966).

The bright red flowers of *Delphinium cardinale* are borne in a horizontal position on tall flower stalks. Stamens and carpels are situated just beneath the entrance to the single horizontally held spur. Pollen adheres to the chins of the hummingbirds which hover in front of these flowers to probe the nectar-containing spurs (Grant and Grant, 1966) (Plate 23).

In many of the western North American hummingbird flowers (Penstemon, Castilleja, Ipomopsis, Zauschneria, Salvia, etc.), nectar is secreted at the base of an elongated calyx or corolla tube, and the hummingbirds probing these flowers receive pollen on the top of the head or upper bill base, and/or the chin or under bill base, depending on whether the anthers are positioned above, below, or symmetrically about the corolla entrance.

For example, the two-lipped tubular flowers of *Mimulus cardinalis* have slightly exserted stamens and stigma positioned above the corolla entrance; pollen is brushed on the tops of the heads of hummingbirds probing them for nectar (Plate 25C–D).

In the bilabiate flowers of *Salvia lemmoni* the two anthers and stigma stand above the corolla tube entrance in the shelter of the upper lip. Hummingbirds contact the reproductive organs with the upper side of the bill as they probe for nectar.

The tubular flowers of *Zauschneria californica latifolia* are borne in a nearly horizontal position; stamens and stigma run along the lower side of the tube and project beyond the corolla entrance so that feeding humming-birds contact these organs with the under sides of their bills and their chins (Plate 26A–B).

The ring of five anthers in the radially symmetrical flowers of *Ipomopsis aggregata* contact the upper or under region of the bill base of the birds probing them for nectar (Plate 24A–B).

Spatial separation of nectar and essential organs is achieved in other species with relatively short floral tubes by elongation of the essential organs. *Trichostema lanatum* (Labiatae) bears spikes of blue flowers typically mint-like in conformation. The extremely long stamens and style arch out from the short corolla tube and brush the tops of the heads of the feeding hummingbirds (Grant and Grant, 1966) (Plates 22, 25A).

Similarly, in *Ribes speciosum* (Saxifragaceae) the stamens and style are long exserted. Here the long stamens of the pendant crimson flowers con-tact the heads or chins of hummingbirds when they probe up from beneath (Grant and Grant, 1966). The exserted stigma is apparently contacted by the bird's chin more often than by its forehead (Plate 26C–D).

Pollen transfer via the bill tips of hummingbirds occurs in some of the plant species with relatively short floral tubes and included reproductive organs.

For example, the honeysuckle, *Lonicera involucrata ledebourii* (Capri-foliaceae) bears tubular yellow flowers heavily tinged with red on the upper side and subtended by bright red bracts. Stamens and stigma are included within the corolla tube. Allen hummingbirds which extensively feed on these flowers contact the reproductive organs with the tips of their bills (Grant and Grant, 1967a).

Bill tip pollination probably occurs accidentally with the visits of hummingbirds to the flowers of manzanitas. In the early spring the western manzanita species bloom out with clusters of delicate white or pale pink short-tubed flowers. Although these Arctostaphylos species (Ericaceae) are extensively visited and probably chiefly pollinated by an array of bees and flies, hummingbirds regularly probe the blossoms for nectar in the early season of the year when few other flowers are available. In California we have observed Black-chinned hummingbirds probing the flowers of *Arctostaphylos glauca eremicola* in the pinyon zone bordering the western edge of the Mojave Desert; Anna, Allen, and Rufous hummingbirds visiting the flowers of *A. parryana pinetorum* in the San Gabriel Mountains; and Calliope hummingbirds probing the tiny flowers of the Pine-mat manzanita (*A. nevadensis*) amidst snow patches in Lassen Volcanic National Park (Grant and Grant, 1966, and unpubl.). In each case some pollination is probably effected by the hummingbirds as their bill tips brush by accident the essential organs within the small urn-shaped flowers (Plate 28D).

Bill tip pollination is in some cases a chance happening and in others probably less effective than transfer of pollen on the feathered body parts of hummingbirds. Not only does a smaller amount of pollen adhere to the bill tip, but, also, hummingbirds often perch and clean their bills by scraping them on bare twigs after a series of flower visits.

Bill tip pollination by hummingbirds can be regarded as a transitional stage in the development of adaptations for hummingbird pollination among the plant species of western North America. We find among these plant species cases where populations are pollinated in part and incidentally by hummingbirds, but are better adapted to pollination by insect visitors (as in Arctostaphylos). In other instances adaptation to hummingbird pollination has reached the racial stage of divergence, as in *Lonicera involucrata ledebourii*.

TABLE 3. PROPORTIONS OF NECTAR-CONTAINING TUBE IN VARIOUS WESTERN AMERICAN HUMMINGBIRD FLOWERS[1]

Species	Family	Length of Floral Tube from Orifice to Base, mm.	Diameter of Floral Tube at Orifice, mm.
Fouquieria splendens	Fouquieriaceae	20	4-5
Ipomoea coccinea	Convolvulaceae	15	2
Ipomopsis aggregata, Sierra Nevada race	Polemoniaceae	15-20	3-4
Salvia spathacea	Labiatae	35	7
Stachys coccinea	Labiatae	20	
Trichostema lanatum	Labiatae	10	1.5
Castilleja stenantha	Scrophulariaceae	15	4
Diplacus puniceus	Scrophulariaceae	35	5
Mimulus cardinalis	Scrophulariaceae	25-30	4
Galvezia speciosa	Scrophulariaceae	16-17	5
Penstemon barbatus	Scrophulariaceae	22	6
Penstemon centranthifolius	Scrophulariaceae	22-25	3
Penstemon newberryi	Scrophulariaceae	25	4-5
Keckia ternata	Scrophulariaceae	25	4
Beloperone californica	Acanthaceae	15	3-4
Bouvardia glaberrima	Rubiaceae	25	2
Lonicera involucrata ledebourii	Caprifoliaceae	15	3
Zauschneria californica latifolia	Onagraceae	22-25	5
Silene laciniata greggii	Caryophyllaceae	25	
Aquilegia formosa truncata	Ranunculaceae	25-30	5
Delphinium cardinale	Ranunculaceae	20-24	2
Delphinium nudicaule	Ranunculaceae	25	2-3

[1] The measurements given are for particular populations studied and do not necessarily represent the range of variation in the whole species.

5 *Cross- and Self-Pollination*

In the preceding chapter some examples were presented of floral mechanisms contributing to the efficiency of pollination among hummingbird flowers.

Hummingbirds follow no regular pattern in the sequence of their flower visits and a feeding trip may include visits to a few flowers on numerous plants or to numerous flowers of one individual plant. This feeding behavior may result in a high frequency of self-pollination among self-compatible species.

Self-incompatibility, protandry, and protogyny are among the various mechanisms promoting cross-pollination of the western North American hummingbird flowers. In the case of individual plant species, however, our present knowledge of the breeding systems and the degree to which cross-pollination is effected by hummingbird pollinators is very scanty.

THE PREVALENCE OF SELF-POLLINATION
IN *Fouquieria splendens*

The Ocotillo, *Fouquieria splendens*, bears clusters of flowers at the tips of long whip-like branches. Numerous yellow stamens are exserted from the bright red tubular corollas. Both stamens and stigma mature simultaneously in this species, and no effective spatial separation of the sexes occurs within the flowers. Plants growing at the Rancho Santa Ana Botanic Garden in Claremont, California, proved to be self-compatible, and although the selfs yielded a relatively low set of capsules and seeds, these seeds produced vigorous seedlings (Grant, 1958). In *Fouquieria splendens*, therefore, viable seeds can be formed by autogamous self-pollination.

34

Furthermore, since the individual plants are fairly widely spaced in the natural populations, and bear numerous flowers each, hummingbirds feed on scores of flowers of one plant before flying to another plant. Their feeding activities thus lead to a preponderance of self-pollinations along with a small proportion of cross-pollinations.

CROSS-POLLINATION IN *Delphinium cardinale*

Self-compatibility is common in the genus Delphinium (Epling and Lewis, 1952), and in *D. cardinale* an interesting mechanism contributing to the efficiency of cross-pollination is superimposed on the protandrous condition.

Epling and Lewis (1952) investigated the population structure and breeding habit of some bee-pollinated Delphinium species in California. In this genus the flowers are borne in spikes and are markedly protandrous. Among the species which are chiefly pollinated by species of Bombus, the lower flowers of the spike mature first, resulting in the lower flowers being in a female condition while the upper flowers are in the male condition. Although self-pollination is possible, cross-pollination is mainly brought about, due to the feeding behavior of the pollinators which systematically visit first the lower flowers of the spike then the upper ones. In this way, the pollen dusted on the bees in their visits to the upper flowers is carried to the lower female stage flowers of the next plant visited.

The related species, *Delphinium cardinale*, is pollinated by hummingbirds which follow no such systematic pattern while feeding on the flowers of an individual spike. A perennial species of rocky foothills and washes of coastal southern California, *D. cardinale* remains dormant throughout the dry season of the year. After the winter rains a rosette of leaves develops from the root crown, followed by one or more tall flowering stalks bearing spikes of red flowers in the late spring.

As the buds open, access to the nectar-secreting spur occurs before the anthers begin to mature. The numerous stamens become erect a few at a time, the anthers of one flower shedding pollen over a period of 6 to 7 days. When the last stamens have matured the three stigma lobes become erect and are receptive for 2 to 3 days.

Small *D. cardinale* plants may bear only a single floral axis, while larget plants normally have in addition secondary, tertiary, or occasionally quaternary floral axes. All of the flowers on one axis open simultaneously and mature at more or less the same rate so that all the flowers on one axis are either in a male or female stage.

As the flowers of the primary axis reach the female stage the flowers on the secondary axis begin to secrete nectar. As the primary-axis flowers drop their petals and the capsules begin to swell, the secondary-axis flowers enter the male stage. The same sequence of maturation takes place with the tertiary and quaternary floral axes, resulting in entire flowering stalks being in either a male or female stage.

As a result of this sequence of floral maturation, those plants which bear a single flowering stalk are at any one time either male or female. Only those larger plants which bear two or more flowering stalks from the root crown may simultaneously have flowers on one stalk in the male condition and flowers on the other in a female condition. Cross-pollination by the hummingbirds is made necessary for young plants, and highly probable for older plants, in this way.

TERRITORIALITY AND POLLEN DISPERSAL IN A POPULATION OF *Delphinium cardinale*

How does the feeding behavior of hummingbirds affect the pattern of pollen dispersal within a population of plants adapted to hummingbird pollination? This problem, of theoretical interest to population biology, is technically difficult to attack.

Identification of both individual plants and individual hummingbirds is in many instances extremely difficult. For this reason, *D. cardinale* was selected for investigation, as individual plants (disregarding products of vegetative reproduction) are fairly easily discernible.

A preliminary study was carried out through observation of the feeding behavior of an adult male Costa holding a territory in a segment of a population of *D. cardinale* growing in a wash in Claremont, California. Although this problem requires much more study, it seems worthwhile to present the generalized scheme of pollen dispersal suggested by the activities of this bird and other birds in its territory.

As large *D. cardinale* plants may bear several flowering stalks branching from the root crown below ground, population size was estimated by counting the number of flowering stalks and dividing it by the average number (1.6) of flowering stalks produced per plant in the population.

Through this method, the size of the *D. cardinale* population was estimated to include 421 plants in late June, 1965. The population was included in an area of about 180 by 150 yards, and the plants grew intermixed with small flowering annuals (*Eriastrum sapphirinum, Centaurium venustum*), and blooming thistles, Opuntia, *Salvia apiana, Dudleya lanceolata,* and *Penstemon spectabilis*. Live oaks and other small trees and shrubs grew at scattered intervals in the area.

Costa hummingbirds were very numerous in the area. Adult males and presumed females and immatures were present (no habitat separation between the sexes occurs in *Calypte costae*, according to Pitelka, 1951). The birds were perching in the small trees and shrubs and darting out to feed on the Delphinium flowers (Plate 23).

The adult males held posts in the trees and defended the surrounding Delphinium plants from intruders. These birds fed mostly on plants in the areas which they defended, but made occasional flights elsewhere to feed. The same post was often held by the same male (judging from the continued use of favorite perching places) for several days or more in succession.

One such adult male Costa which was observed at a post in a small Elderberry tree for over a week was selected as the subject of observation. This bird's feeding territory was determined as including those plants which it both regularly fed on and defended from intruding birds. About 40 flowering stalks or 25 *D. cardinale* plants were included in this area.

The activities of this bird within its territory and the activities of intruding hummingbirds (including other adult males, females, immatures, and rarely a male Allen) were observed and recorded over 3- and 3½-hour periods on two successive mornings (July 3, 7:15-10:15 A.M.; July 4, 5:30-9:00 A.M.).

During this 6½-hour period the male Costa made 42 feeding trips, visiting an average of 31.2 flowers per trip and a total of 1,311 flowers. Of these 42 feeding trips, 37 (88 per cent) were within the territory defended by the bird, while 5 (12 per cent) were outside the defended area.

During the same period 12 feeding trips were made by intruding hummingbirds into this territory. These intruders visited an average of 16 flowers per trip, or a total of 192 flowers. The much lower number of flowers visited per feeding trip by intruders is no doubt due to their being chased away by the male Costa.

No quantitative expression of pollen dispersal can be made from these figures, but the general pattern is clear. The territorial activities of this male Costa resulted in pollen dispersal largely within the area defended by the bird. The occasional far-flung flights of the male Costa must have resulted not only in pollen being transported to distant Delphinium plants, but also in the introduction of foreign pollen into the home territory. Similar results must have been brought about by the feeding visits of intruding birds.

In June 1966, following a very dry winter, the same population of *Delphinium cardinale* was much decimated in size, the number of flowering plants being estimated as 134. Although Costa hummingbirds were regular visitors to the flowers, they were much fewer in number than in the preceding year.

The pattern of pollen dispersal within a segment of the population proved very different from that of 1965. Despite daily observation of the population, no adult males were observed to hold feeding territories for successive days. Instead, the feeding activities of the hummingbirds in the Delphinium population appeared to be more or less random with respect to the sequence of plants visited.

These observations indicate that patterns of pollen dispersal brought about by hummingbirds may vary widely from year to year in the same plant population. In 1965 the larger size of the plant population supported numerous hummingbirds, and the closely grouped plants were easily defended by males as feeding territories. In 1966 unfavorable growing conditions resulted in few and widely scattered plants, providing food for fewer birds. Feeding territories were not held by the males, perhaps because an adequate supply of flowers was distributed over too great an area to be

38

defended, and pollen was therefore dispersed by the birds in a more random fashion throughout the plant population.

We can conclude that the two distinct patterns of pollen dispersal in this population of *D. cardinale* suggested by distinct differences in the feeding behavior of the hummingbirds in two successive years are a result of the complex interactions between the components of this ecological system. Favorable or unfavorable growing conditions affect the size and structure of the plant population. This in turn is a factor determining the numbers of hummingbird pollinators in the plant population and their patterns of feeding behavior.

6 Geographical Distribution of Hummingbird Flowers in Western North America

Hummingbird flowers occur widely throughout western North America. They vary in frequency, however, from one region to another within this large sub-continental area. In this chapter we shall examine some trends in their geographical distribution.

The floras of large political subdivisions—states or provinces—provide a useful basis for comparing the abundance of hummingbird-pollinated plant species in different regions of western North America, and hence for examining geographical trends in this assemblage of plants.

We checked several representative state floras for the presence of the hummingbird-pollinated plant species listed in Chapter 3. In making the tabulations we have attempted to correct for differences in nomenclature between different state floras. The floras used for six western states were: California (Jepson, 1925; Munz and Keck, 1959); Arizona (Kearney and Peebles, 1960); New Mexico (Wooton and Standley, 1915); Colorado (Harrington, 1954); Idaho (Davis, 1952); and Oregon (Peck, 1961). The floras of two northwestern regions were also scanned, namely British Columbia (Henry, 1915) and Alaska (Hultén, 1941-1949; Anderson, 1959; Wiggins and Thomas, 1962).

SPECIES DENSITY IN WESTERN STATES

The estimates of the numbers of hummingbird-pollinated plant species per state are given in Table 4. It will be recalled that, on account of the policy adopted in compiling the general list, these are conservative estimates. We see that California, with an estimated 80 species of hummingbird flowers.

TABLE 4. SPECIES DENSITY OF HUMMINGBIRD FLOWERS IN SEVERAL WESTERN STATES

State	No. Species in State	No. Species per 100,000 Sq. Mi.
California	80	51.0
Arizona	50	44.0
New Mexico	40	32.9
Colorado	19	18.3
Idaho	14	16.9
Oregon	28	29.1
British Columbia, southern half	10	5.6
Alaska	5	0.9

has the greatest number. Arizona and New Mexico follow with 50 and 40 species respectively. The numbers then drop off to the north in Oregon and in the Rocky Mountain region. Fewer species of hummingbird-pollinated plants occur in southern British Columbia (10) and still fewer in Alaska (5).

If we correct for differences in size between the states, by computing the number of hummingbird-pollinated species per 100,000 square miles of land area, we obtain more directly comparable figures. Such figures are given in the righthand column of Table 4 and are shown graphically in Map 3. California still has the highest concentration of hummingbird flowers. But an area of equal size in Arizona has almost as high a concentration, the differential between the two states being reduced.

GEOGRAPHICAL GRADIENTS

The Pacific slope appears to be richer in hummingbird-pollinated species than areas of equal size and latitude in the Rocky Mountain region. The species density shows a regular decline to the north, however, along both the Pacific and the Rocky Mountain axes.

The north-south gradient in numbers of hummingbird-pollinated plant species can be observed within smaller distances when a state area is subdivided into regions and the floras of these regions are considered separately. Thus in Arizona there are 34 species of hummingbird flowers in the southern third of the state as compared with 25 species in the northern third. The

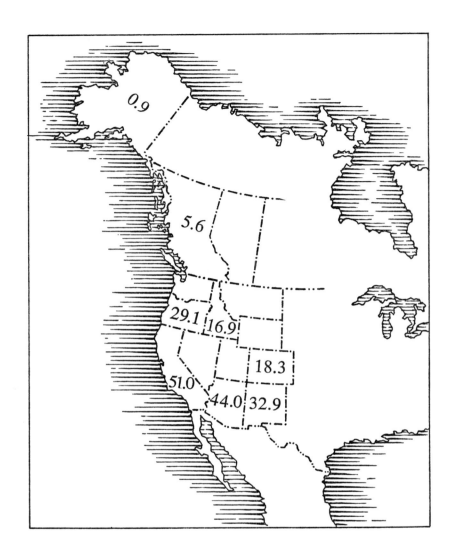

MAP 3.   NUMBER OF SPECIES OF HUMMINGBIRD FLOWERS PER 100,000 SQUARE MILES
OF LAND AREA IN VARIOUS WESTERN AND NORTHWESTERN STATES

The figure for British Columbia refers to the southern half of that province.

following species drop out in passing from southern to central Arizona:

Erythrina flabelliformis	Penstemon lanceolatus
Polemonium pauciflorum	Penstemon parryi
Salvia henryi	Penstemon pinifolius
Salvia lemmoni	Jacobinia ovata
Castilleja austromontana	Beloperone californica
Castilleja cruenta	Bouvardia glaberrima
Castilleja lanata	Lobelia laxiflora
Castilleja laxa	Agave schottii
Castilleja patriotica	

Similarly, on the Pacific slope, a group of bird-pollinated species range through northern California to the southwestern corner of Oregon, but then drop out between southern and central Oregon. Species of hummingbird flowers which have this distribution pattern are:

Silene californica	Pedicularis densiflora
Delphinium nudicaule	Brodiaea ida-maia
Zauschneria californica	Lilium parvum
Mimulus cardinalis	Fritillaria recurva
Diplacus aurantiacus	

Proceeding north, another zone in which bird-pollinated species drop out of the flora lies between southern British Columbia and southeastern Alaska. Ten species of known or probable hummingbird flowers occur in southern British Columbia. All ten range farther south, at least to Oregon. But seven of them reach their northern limit in southern British Columbia. The seven species in question are:

Ipomopsis aggregata	Castilleja rupicola
Stachys ciliata	Castilleja suksdorfii
Castilleja angustifolia	Lonicera ciliosa
Castilleja rhexifolia	

In Alaska, finally, we find only five plant species with floral characters suggesting hummingbird pollination, namely *Aquilegia formosa* and four Castillejas. These are all confined to southern Alaska from the southeastern coast to the Kenai Peninsula. Hummingbird flowers disappear completely

43

from the flora as we proceed north of the Kenai Peninsula into central and northern Alaska.

The decline in abundance of bird-pollinated plant species at higher latitudes is correlated with the similar decline in the abundance of hummingbirds described in Chapter 2. The great abundance of hummingbird individuals in the southern areas is undoubtedly the primary factor responsible for the occurrence of relatively numerous species of bird flowers in the same areas. The decrease in numbers of hummingbirds to the north is thus not only correlated with, but probably determines, the decrease in numbers of bird-pollinated plant species, up to the northernmost limits of both parties in southeastern Alaska.

7 Ecological Distribution of Hummingbird Flowers in Western North America

The hummingbird flowers of western North America occur in a wide variety of ecological habitats, including desert, arid woodland, mountain forests, and coastal and arid scrub communities. They are notably absent in several other plant communities. In the present chapter we will consider some plant communities which include hummingbirds and hummingbird flowers. Several transects through different regions of California will illustrate the absence of hummingbird flowers in certain habitats. Finally, we will consider the correlation between the ecological distribution of hummingbird flowers and hummingbirds in western North America.

THE COLORADO DESERT

After the winter season of rainfall, desert shrubs renew their growth and small annuals rapidly germinate, grow, and flower before the onset of the hot and dry summer (Plate 1A–B).

On the Colorado Desert in California (Map 4) early spring-blooming annuals cover the bare and rocky mountain slopes and desert basins. On lower mountain slopes and desert mesas, the Ocotillo (*Fouquieria splendens*), a widespread species in the arid Southwest, forms large and dense stands (Plate 1A). The numerous tall straight shoots of this perennial species are clothed with long spines and new small green and fleshy leaves. In late winter and early spring clusters of bright red waxy flowers bloom out at the tips of the tall unbranched shoots.

The dense stands of *Fouquieria splendens* which often stretch for many miles provide the chief source of nectar for the hummingbirds which migrate

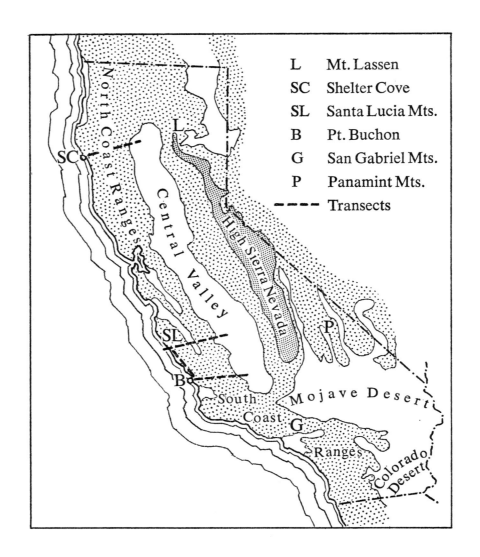

L   Mt. Lassen
SC   Shelter Cove
SL   Santa Lucia Mts.
B   Pt. Buchon
G   San Gabriel Mts.
P   Panamint Mts.
- - - -   Transects

MAP 4. MAP OF CALIFORNIA SHOWING PHYSIOGRAPHIC PROVINCES, LOCALITIES, AND TRANSECTS MENTIONED IN THE TEXT

through this region to their breeding ranges to the north, as well as for Costa hummingbirds which breed in desert regions of southeastern California and southwestern Arizona.

We have watched Costa, Rufous, and unidentified hummingbirds probing the flowers of *Fouquieria splendens* in different years and at various localities on the Colorado Desert. Linnets and migrating Western tanagers also visit the flowers.

Secondary in abundance to the Ocotillo, but starting to bloom earlier in the season, the Chuparosa (*Beloperone californica*) typically grows along desert washes and water courses at generally lower elevations than *Fouquieria splendens* (Plate 1B). Before the leaves have emerged, numerous tubular dark red flowers cover the bare twigs of this perennial shrub. On warm sunny days from early spring into May hummingbirds may be seen in large numbers darting about the blooming Beloperone bushes, feeding on the flowers, and chasing one another.

Honeybees also gather nectar from the flowers of *Beloperone californica*, and Linnets and Gambel's sparrows bite off and eat the nectar-filled flower bases (Jaeger, 1940).

Costa hummingbirds may winter on the Colorado Desert (Bent, 1940; Grinnell and Miller, 1944), and migrants may arrive there before *Beloperone californica* and *Fouquieria splendens* begin to flower. At this season they are probably more or less dependent on insect food. Hummingbirds also occur during the flowering season in areas lacking in hummingbird flowers; in such situations the birds commonly utilize any flowers which provide a nectar source, as, for example, *Isomeris arborea, Hyptis emoryi, Phacelia minor,* and *Lotus rigidus.* Grinnell and Miller (1944) report *Chilopsis linearis* and species of Lycium as much visited by *Calypte costae*.

PINYON-JUNIPER WOODLAND

At higher elevations bordering the California deserts and in arid regions of southern California Pinyon-juniper woodland occurs as an open stand of small trees and shrubs. Hummingbird flowers growing in this arid formation include various species of Castilleja, *Astragalus coccineus*, and various Penstemons.

Hummingbird flowers are absent from the Mojave Desert but appear in the Pinyon-juniper woodland of the higher mountain ranges rising above and bordering the desert. To the west of Death Valley the Panamint Mountains rise abruptly from the valley floor (Map 4). Above ca. 4000 feet, from the upper desert slopes into the pinyon-juniper woodland, *Castilleja chromosa* blooms out with bright red flowers in late spring.

COASTAL SAGE SCRUB AND CHAPARRAL OF SOUTHERN CALIFORNIA

The coastal sage scrub, characteristic of the dry rocky coastal plain of southern California, is composed of low woody shrubs such as *Artemesia californica*, *Rhus integrifolia*, *Erigonum fasciculatum*, *Salvia apiana*, and *Salvia mellifera* (Plate 2C). At higher elevations the coastal sage scrub intergrades with the chaparral, a denser vegetation of somewhat taller shrubs and small trees occurring on dry rocky slopes and ridges (Plate 2B). Chaparral species, such as *Adenostema fasciculatum*, *Quercus dumosa*, *Yucca whipplei*, *Fremontia californica*, and Rhamnus, Ceanothus, and Arctostaphylos species, are adapted to rapid growth during the mild and rainy winter season, and dormancy during the long hot dry summer.

In these communities conspicuous hummingbird flowers include *Delphinium cardinale*, *Diplacus puniceus*, *Silene laciniata*, and Castilleja.

In the rock-strewn washes on the south side of the San Gabriel Mountains (Map 4), Costa hummingbirds are frequent visitors to *Delphinium cardinale*. *Calypte anna* is more at home in the chaparral zone where the females commonly nest in live oaks in canyon bottoms, and the males establish territories on the more open hillsides and canyon walls (Grinnell and Miller, 1944).

In spring, at lower elevations in the San Gabriel Mountains, Anna hummingbirds feed on *Castilleja martinii*, *C. foliolosa*, and *Diplacus longiflorus*. The latter species is also much visited by the long-tongued cyrtid fly, *Eulonchus smaragdinus* (Beeks, 1962; Grant, 1966).

THE COAST RANGES OF CALIFORNIA

Hummingbird flowers bloom throughout the spring in the coastal hills of southern California. *Ribes speciosum* grows in shaded canyons and bears bright red pendant flowers in early spring; this species is much visited by Anna hummingbirds in southern California (Bent, 1940; Grant and Grant, 1966). *Salvia spathacea* produces whorls of dark red flowers in late spring. Both Allen and Black-chinned hummingbirds visit the flowers of this species in the Santa Barbara region in early May (Dawson, 1923).

At Pt. Buchon on the Pacific coast of south-central California (Map 4), the Allen hummingbird breeds in the dense willow thickets along canyon bottom streams. Honeysuckle (*Lonicera involucrata ledebourii*) occurs beneath the willows in the tangled undergrowth of Blackberries and Poison oak, and Allen hummingbirds extensively visit these tubular yellow and red flowers (Grant and Grant, 1967a).

Just above the canyon bottom streams, on dry brush-covered hillsides, Anna hummingbirds are abundant and feed on the flowers of chaparral plant species—*Castilleja sp.*, *Silene laciniata*, and *Diplacus aurantiacus*. Some overlap in the feeding areas of Anna and Allen hummingbirds occur; however, differences in the ecological habitat preferences of breeding Anna and Allen hummingbirds (Pitelka, 1951) result in the pollination of these plant species predominantly by one or the other hummingbird species.

Hummingbird flowers occur only in scattered localities along a transect from Pt. Buchon east through the inner South Coast Ranges, which become progressively more arid toward the interior. *Trichostema lanatum* occurs locally on chaparral-covered slopes (Plate 2B) and *Penstemon centranthifolius* in the dry Digger pine and Pinyon-juniper zones (Plate 2A). The large intervening areas of oak savannah and grassland are devoid of hummingbird flowers.

Hummingbird flowers are absent along the coastal plain of low grass-covered hills ranging from Pt. Buchon north to the Santa Lucia Mountains (Map 4). At the southern tip of this mountain range hummingbird flowers again appear. On the coastal bluffs and in forested canyons, species of Castilleja and *Aquilegia formosa truncata* bloom in late spring (Plate 3B). *Delphinium nudicaule* grows on shaded hillsides in the mixed deciduous forest to the interior. On the arid east slopes of the Santa Lucia Mountains *Penstemon centranthifolius* again appears. Hummingbird flowers are once more absent in the inner hills and valleys with oak savannah and grass vegetation.

Penstemon centranthifolius has an interesting distribution. This species occurs in a narrow zone of arid foothills and badlands in the South Coast Range of central California and the San Gabriel and other ranges of southern California (Map 4). It is characteristically found on the interior side of these ranges, and approaches the coast in only a few exceptional places. In the South Coast Range it usually occurs in Digger pine woodland (Plate 2A), in the San Gabriel and adjacent ranges in Pinyon-juniper woodland, and farther south in various arid scrub formations (Plate 1C). Its distribution area has the form of a continuous or nearly continuous, but branched, strip several miles wide and at least 300 miles long.

Penstemon centranthifolius blooms in late spring, and furnishes the dominant and often the only hummingbird flower in its arid inner foothill strip. The plant communities adjacent to this strip on both sides, moreover, often have no hummingbird flowers at all, as mentioned previously.

Costa hummingbirds feed extensively on *Penstemon centranthifolius* in southern and central California. As a result of the distribution of this plant in dry open habitats favored by *Calypte costae*, the Penstemon is probably pollinated predominantly by the Costa. Conversely, the latter probably relies mainly on this Penstemon for food in the many parts of its range where few or no other hummingbird flowers are found.

At Shelter Cove, in coastal northern California (Map 4), Castillejas and *Diplacus aurantiacus* bloom in early summer on the steep coastal bluffs (Plate 3A). *Silene californica* occurs at the edge of the dense forest of Lithocarpus, Madrone, and conifers. Toward the interior these forest-covered ridges are replaced by dense redwood groves, and here no hummingbird flowers occur in the deep shade and litter of the redwood forest. *Mimulus cardinalis* and *Keckia corymbosa* occur in openings in the redwood forest, along river banks, and with *Brodiaea ida-maia* and *Silene californica* in the more open interior forest to the east.

THE HIGH MOUNTAINS

At the southern tip of the Cascade Range, Mt. Lassen rises to a height of over 10,000 feet (Map 4, Place 4C). This peak and the adjacent high peaks are still covered with snow patches in mid summer. At these high elevations Calliope hummingbirds were very numerous in mid July, 1966.

In open spots in the coniferous forest Calliope hummingbirds probed the red trumpet-shaped flowers of *Ipomopsis aggregata*, and the half-opened flowers of a *Castilleja*. On rocky ledges above cold mountain streams bordered with early spring flowers, *Penstemon newberryi* and *Castilleja miniata* were much visited by these hummingbirds.

At higher elevations green mats of *Arctostaphylos nevadensis* occurred amidst the snow patches beneath the conifers. Clusters of tiny white flowers, each secreting a large drop of nectar, hung beneath the glossy green foliage. Many Calliope hummingbirds and a single male Rufous probed these flowers on a cold and windy morning.

This situation in which the birds had arrived in an area prior to the opening of the hummingbird-flowered species was observed again on the lower slopes of Lassen Peak. *Castilleja payneae* was abundant on the steep rocky slopes, its rusty red flowers still in bud. Numerous Calliope hummingbirds perched in the stunted White bark pines at timberline, flying down to probe the yet unopened flowers.

Numerous hummingbird flowers bloom together in mid and late summer in the high montane forests of the Sierra Nevada (Map 4, Plate 4B). For example, common species occurring to the east of the Sierran Crest are:

Aquilegia formosa	Castilleja breweri
Ipomopsis aggregata	Castilleja linariaefolia
Penstemon bridgesii	Castilleja miniata
Penstemon newberryi	

The flowers of these plant species are regularly visited by Calliope hummingbirds breeding in these mountain forests, and by southward migrants, especially the Rufous, and less frequently, the Allen.

The feeding flights of the Calliopes extend nearly to timberline with the altitudinal limits of *Penstemon newberryi*, and occasionally above, for they were observed to visit the flowers of an introgressive population of *Penstemon menziesii davidsonii* on an alpine fell field at 11,000 feet elevation.

At one locality hummingbirds were systematically feeding on the flowers

51

of *Castilleja miniata*, *C. breweri*, *Penstemon bridgesii*, and *Aquilegia formosa*, while completely neglecting those of *Ipomopsis aggregata*. A check of the nectar yield of these species on several successive days revealed that those species visited by the birds produced abundant nectar, while *I. aggregata* yielded a small quantity in the early morning, but by mid morning had no detectable nectar. These observations suggest that nectar yield is an important factor in the competition between plant species for pollinating visits by hummingbirds.

In the high mountains of southern California hummingbird flowers bloom from mid summer into late October. At higher elevations in the San Gabriel Mountains (Map 4), the following species make a conspicuous display in the dry summer season when few other plant species are in flower.

Aquilegia formosa	Keckia ternata
Mimulus cardinalis	Zauschneria californica latifolia
Penstemon bridgesii	

Anna hummingbirds, abundant at lower elevations in the same general region during winter and spring, are very numerous in these mountains in summer and fall, feeding on the abundant hummingbird flowers. Many individuals of this species remain at high elevations, feeding on *Zauschneria californica latifolia* into October and descending to lower elevations with the first storms of fall. A few Annas remain through the first snows and subfreezing temperatures of early winter.

Southward migrants passing through these mountains include Calliope, Rufous, and Allen hummingbirds. In late summer all four species may be seen feeding together on the especially abundant *Zauschneria californica latifolia*.

The unidentified hummingbird shown in Plate 21 defended a patch of Zauschneria flowers, not only from other hummingbirds, but from an intruding carpenter bee as well. A large black bee (Xylocopa sp.) entered the Zauschneria patch and went from flower to flower, piercing the corolla tubes to collect the nectar. At intervals the hummingbird attacked the bee, the collision between the two making a noise like that of inserting a flexible stick into the moving blades of a fan. With each attack (and perhaps

20 occurred in less than an hour) the bee would retreat slightly, or fly to some flowers a few feet away, but persisted in its stealing of nectar despite the attacks of the bird.

Ipomopsis aggregata occurs in great abundance in openings of the coniferous forest in the high (6500-7500 feet) plateau country of north central Arizona (Plate 4A). At higher elevations, as in the Aspen-conifer forest of the San Francisco Peaks, *Penstemon barbatus* replaces *Ipomopsis aggregata* as the most common hummingbird flower. At very high elevations in these mountains (up to about 11,000 feet) *Castilleja miniata* grows in openings in the forest up to timberline. Broad-tailed and Rufous hummingbirds are numerous visitors to these hummingbird flowers in late summer.

In the Chiricahua Mountains of southeastern Arizona mid and late summer-blooming hummingbird flowers occur mostly above 6000 feet and abundantly at elevations above *ca.* 8000 feet. Hummingbird species in this region include the Black-chin, abundant at lower elevations, the Blue-throat, and the Rivoli, Calliope, Broad-tail, and Rufous at higher elevations.

In the coniferous forest at higher elevations in these mountains, hummingbird flowers include:

Aquilegia triternata	Castilleja patriotica
Silene laciniata greggii	Salvia lemmoni
Polemonium pauciflorum	Penstemon barbatus

Salvia lemmoni, common on rocky slopes in openings in the forest, bears spikes of bright rose-colored flowers. On a sunny morning in late August, 1961, four species of hummingbirds (the Rufous, Black-chin, Broad-tail, and Rivoli) fed together on the Salvia flowers. The Salvia pollen was conspicuous as a yellow patch on the upper side of the bill of those birds frequenting the flowers. In the partial shade of the forest, the long spikes of brilliant red *Penstemon barbatus* flowers were much visited by Black-chinned, Rufous, Rivoli, Calliope, and Broad-tailed hummingbirds.

CORRELATION BETWEEN THE ECOLOGICAL DISTRIBUTION
OF HUMMINGBIRD FLOWERS AND HUMMINGBIRDS

As we have seen in the foregoing pages, the western North American hummingbird flowers are often visited by several species of hummingbirds. Due to the similarity in bill length and shape among the western North American hummingbirds, any of these species can effect pollination of these hummingbird flowers. In regions inhabited by two or more breeding hummingbird species, ecological habitat preferences of the different hummingbird species are the chief factor involved in those cases where a plant species is pollinated mainly by one hummingbird species. During the spring and fall migration seasons the ecological barrier between the breeding hummingbird species breaks down, and several hummingbird species commonly pollinate the same hummingbird flowers blooming together at this time.

In western North America hummingbird flowers are absent from grassland and savannah country, in the deep shade of dense forests, and in the alpine zone of the high mountains. Although they occur on the Colorado Desert they are absent from the Mojave Desert.

The absence of hummingbird flowers in these ecological habitats is paralleled by the absence of breeding hummingbirds in the same habitats. Miller (1951) records in addition to seashore, marsh, and aquatic habitats, those of sagebrush formations, savannah, grassland, and alpine meadow habitats as lacking in breeding hummingbird species.

Hummingbird species are completely lacking in the Great Plains region of North America, suggesting that the ecological factors barring them from savannah and grassland country in western North America may be an extension of their general absence from this type of terrain.

Lack of nesting sites and the extremely short warm season in alpine zones may be the chief factors barring hummingbirds from these habitats. The factors barring them from the Mojave Desert remain unknown.

In summary, we find hummingbirds and hummingbird flowers occurring together in a wide variety of ecological habitats in western North America. In certain other habitats both hummingbird flowers and breeding hummingbird species are lacking, and the factors excluding them from these habitats remain unexplained.

Hummingbird flowers are found only in those habitats which also include breeding hummingbird species. This correlation suggests that hummingbirds have been able to exert an effective selective pressure on the vegetation—a pressure leading to the development of floral adaptations for bird pollination—only in communities where the birds have been a factor continuously for some length of time.

8 *Origins of Western American Bird Flowers*

The list of known or probable bird flowers presented in Chapter 3 reveals, on analysis, several tendencies in life form, floral form, and systematic relationships. These tendencies provide significant clues as to the conditions favoring the evolutionary development of the hummingbird pollination system in western North America.

LIFE FORM

The overwhelming preponderance of the species on the list of bird flowers are either perennial herbs (such as Castilleja, Penstemon, Aquilegia) or soft-wood subshrubs, (i.e., Diplacus, Keckia, Beloperone).

True shrubs with hard wood are represented less commonly among our bird-pollinated plants. The only genera exhibiting this life form are Erythrina, Ribes, Fouquieria, and Lonicera.

Annual herbs are exceptional among western bird-flowered plants. So far as we can determine there are only five annual species in the whole assemblage of 129 species. These are:

Ipomoea coccinea	Castilleja stenantha
Gilia splendens	Castilleja exilis
	Castilleja minor

A perennial life cycle is evidently a favorable pre-condition, or pre-adaptation, for the development of hummingbird flowers in western North America. The reason for this can be suggested. Hummingbirds consume great quantities of nectar over prolonged periods, especially in their breeding and nesting territories. Plants which are most successful as hummingbird flowers, therefore, are probably those which can produce a continuous succession of flowers over a fairly long blooming season. Perennial herbs, subshrubs, and shrubs can do this, whereas annual herbs with their relatively short flowering season usually cannot.

COROLLA STRUCTURE

The sympetalous Dicotyledons have by far the largest representation on the list of any main subdivision of Angiosperms, with 107 of the 129 species. The non-sympetalous Dicotyledons and Monocotyledons contribute 15 and 7 species respectively.

These figures deviate markedly from the proportionate representation of the three sub-classes of Angiosperms in the flora as a whole. This can be shown by an analysis of the California flora, where the proportions are very nearly the same as those in the west generally. The sympetalous Dicotyledons constitute an estimated 39 per cent of the California flora, but contribute 78 per cent of the bird-pollinated species. By comparison, the apetalous and choripetalous Dicotyledons, with 43 per cent of the species in California, have only 16 per cent of the hummingbird flowers.

The explanation of these facts is not far to seek. The sympetalous corolla with its tubular construction is another favorable precondition for the evolutionary development of flowers adapted to hold and offer nectar to long-billed hummingbirds.

This explanation is confirmed by the observation that the non-sympetalous bird flowers on our list accomplish the same end result—a tubular shape—in some alternative way. In Silene a floral tube is formed by a tubular calyx which holds the free petals together; in Aquilegia by the development of long spurs on the petals; in Echinocereus by numerous perianth segments imbricated along an elongated hypanthium.

BEE POLLINATION AS AN ANCESTRAL CONDITION

Let us consider next the types of pollination system which are ancestral to hummingbird pollination in the western American flora. On this as on any other phylogenetic question we cannot expect to obtain absolutely certain answers. We do, however, find some highly suggestive indications from a survey of the modes of pollination prevailing in the closest non-bird-pollinated relatives of our hummingbird flowers.

The information on pollination systems in various plant groups which is needed for our present purpose has been taken from several sources. Much of this information is summarized in the great compendious work of Knuth (1906-1909) and Loew (1904-1905). Many old observations are brought up to date and new ones added by Faegri and van der Pijl (1966). The mode of pollination in various western American plant groups has been worked out by our students and ourselves. The studies of the following groups are particularly relevant here: Aquilegia (Grant, 1952); Penstemon (Straw, 1956a, 1956b); Lilium (Davis, 1956); Pedicularis (Sprague, 1962); Scrophularia (Shaw, 1962); Salvia (Grant and Grant 1964); and Polemoniaceae (Grant and Grant, 1965). Finally, we have been able to draw on a considerable body of unpublished observations of our own.

A comparative survey of modes of pollination within various plant genera, based on the several bodies of evidence mentioned above, suggests strongly that hummingbird flowers are derived from bee flowers in numerous independent phyletic lines in the western American flora. This hypothesis has been put forward earlier (Grant, 1961; Grant and Grant, 1965), and is developed further here.

The genus Aquilegia (Ranunculaceae) has its most primitive forms and its center of distribution in Eurasia. Bee pollination prevails in the Eurasian species and in a few related species in cold-temperate regions of North America. The bird-pollinated species—*Aquilegia canadensis* in eastern North America, and A. *formosa* and others in the West—are related to and apparently derived from the bee-pollinated species.

The most probable phylogeny is that bee-pollinated Aquilegias migrated from their Eurasian center of distribution into North America and there became exposed for the first time in the history of the genus to hummingbird visitations. Some of these early immigrant Aquilegias then diverged from the ancestral pollination system and became adapted, by natural selection, to a different type of pollinator in their new territory, namely hummingbirds (Grant, 1952).

A similar pattern recurs in the genus Delphinium of the same family Ranunculaceae. This widespread north-temperate genus consists almost exclusively of bee flowers. But two species in California and Oregon have bird flowers. These species—*Delphinium cardinale* and *D. nudicaule*—must have diverged from bee-flowered ancestors.

In the family Scrophulariaceae, Pedicularis is another large genus, primarily northern and primarily bee-pollinated, which has given rise to a bird-pollinated species in California, *Pedicularis densiflora* (Sprague, 1962). Similarly, the genus Scrophularia, which is predominantly bee-pollinated and also to some extent wasp-pollinated, has produced one species with hummingbird flowers, *S. coccinea*, in New Mexico (Shaw, 1962).

In Penstemon (Scrophulariaceae) bee pollination is clearly the basic condition and hummingbird pollination derived (Pennell, 1935; Straw, 1956a, 1956b). Here the change-over from bee to bird pollination has occurred, not once, but repeatedly in several different sections of the genus. *Penstemon centranthifolius* and *P. utahensis* are closely related to bee-pollinated species in one section; *P. newberryi* and *P. rupicola* to bee-flowered species in another section; and two or three other sections also contain bird-pollinated species.

The large genus Astragalus (Papilionaceae) with its bee flowers has a single species in the southwestern desert which bears hummingbird flowers, *A. coccineus*. Salvia (Labiatae), a genus which furnishes some of the classical examples of bee flowers, is represented by at least three species of bird flowers in our area, and by others in other parts of the American hemisphere. Polemonium (Polemoniaceae), another bee-flowered genus, is represented by the bird flowered *P. pauciflorum* in southern Arizona (Grant and Grant, 1965).

The pattern just described is found in the 19 genera listed below. In each genus bee pollination is a widespread and evidently original condition. But each genus has given rise to one or a few bird-flowered species in western North America.

Aquilegia	Polemonium	Trichostema
Delphinium	Collomia	Mimulus
Ribes	Satureja	Pedicularis

Astragalus	Monardella	Penstemon
Gilia	Salvia	Scrophularia
Ipomopsis	Stachys	Lonicera
		Brodiaea

The same pattern is displayed again at a higher taxonomic level in three groups of genera.

Let us compare the related genera Epilobium and Zauschneria (Onagraceae). Epilobium is a large, geographically widespread, predominantly bee-pollinated genus. Zauschneria is a minor genus of four species in western North America, all of which have hummingbird flowers. Zauschneria is probably derived from some ancestral form in Epilobium.

Similarly in the Scrophulariaceae, the minor genus Diplacus, which is confined to the Pacific coastal region of North America and is bird-pollinated to a large extent, is probably derived from the large, widely distributed, predominantly bee-flowered genus Mimulus. And the minor bird-flowered genus Galvezia represents a specialized offshoot of the basically bee-pollinated Antirrhinum tribe of the same family Scrophulariaceae.

The 22 phyletic lines which have changed over from bee to hummingbird pollination in western North America are all plant groups with north-temperate floristic affinities. It will be recalled that the hummingbirds are essentially a tropical and subtropical family. These two generalizations, when put together, suggest that the change-over in mode of pollination is a consequence of northern floristic elements coming into contact with subtropical hummingbirds in temperate North America during the course of plant migrations and bird migrations.

LEPIDOPTERA POLLINATION AS AN ANCESTRAL CONDITION

The genera Silene and Lilium have given rise in western North America to two species of hummingbird flowers each. In Silene these species have well-marked bird flowers, and in Lilium their specializations are less well developed. Silene and Lilium are large north-temperate genera in which Lepidoptera pollination prevails. Here, therefore, the change-over has been from Lepidoptera to hummingbird pollination.

These two cases show that northern plants pollinated by agents other

60

than bees, on coming into contact with hummingbirds, can likewise produce bird flowers in some divergent species. But such change-overs have occurred much less frequently than that from bee to bird pollination.

BIRD POLLINATION AS THE ANCESTRAL CONDITION

There is another contingent of hummingbird flowers in western North America which does not conform to the patterns described above. This contingent consists of:

Erythrina flabelliformis	Beloperone californica
Fouquieria splendens	Jacobinia ovata
Anisacanthus thurberi	Bouvardia glaberrima.

These species have several features in common. They are all found in the warm southern part of our area, and extend south into northern Mexico at least. The genera they belong to all occur in subtropical America and some of them also in the American tropics. Finally, the genera involved also have other more southern species which bear apparent bird flowers.

There is no reason to postulate any change-over in mode of pollination during the evolution of these species. The simplest interpretation of the facts is that the species in question, or their immediate ancestors, have followed the hummingbirds, their ancestral pollinating agents, to the areas they now occupy.

CONDITIONS FAVORING THE SHIFT
FROM BEE TO BIRD POLLINATION

The evolutionary development of bird flowers from bee flowers presupposes that hummingbirds visit bee flowers with some frequency and pollinate them with some degree of effectiveness. These preconditions are realized in the modern western flora and undoubtedly existed in the historical past.

Hummingbirds feed regularly on bee-flowered species of Arctostaphylos, Lathyrus, Oenothera, Isomeris, Polemonium, Phacelia, Penstemon, Salvia, Monardella, and other genera (Plates 28-30). Such bee flowers afford nectar in a tube or false tube where it can be extracted successfully by the birds.

Therefore the birds return again and again to these flowers. The illegitimate visits lead to at least some pollination in most of the genera mentioned above.

One gains the general impression from much time spent in the field with hummingbirds that these animals feed on bee flowers more frequently by far than on any other type of non-bird flower. Actual data bearing on this question are greatly to be desired.

The general class of bee flowers includes a wide range of forms as regards degree of specialization for bee visits. At one end of the spectrum we have open tubular flowers, as found for example in Penstemon. At the other end are two-lipped and banner flowers with more or less concealed nectar and reproductive organs, and often with a closed orifice, such as are found frequently in the Papilionaceae and Labiatae. Here the floral mechanism excludes most types of potential visitors except strong or heavy bees.

The various types of bee flowers do not appear to have equal potentialities for making the evolutionary transformation to hummingbird flowers. This transformation is probably difficult in the case of bee flowers which have evolved far down the path of specialization for strong or heavy bees.

It is noteworthy that the Papilionaceae and Labiatae, though represented by numerous species in western North America, have produced relatively few species of hummingbird flowers. Our list (Chapter 3) shows two bird-flowered species in the Papilionaceae and ten in the Labiatae.

By contrast, the single family Scrophulariaceae is represented by an estimated 74 species of bird-pollinated plants in the same western American area. Within the California flora, 15 per cent of the species of Scrophulariaceae have hummingbird flowers; whereas the corresponding proportions for the Labiatae and Papilionaceae in the same region are 5 per cent and 0.3 per cent respectively.

Several different factors probably contribute to the success with which the Scrophulariaceae have been able to evolve hummingbird flowers in western America. These factors are not all known. But one important pre-adaptive condition in the Scrophulariaceae may well be the lack of extreme specialization for bees in most bee-flowered members of this family.

The southern contingent of hummingbird flowers in western North America has its phylogenetic roots in the American subtropics and tropics—in the ancient homeland of the Trochilidae. The interrelationships between the southern bird-pollinated plant species or their ancestors and the humming-birds must have existed for a long time. This long-standing relationship is reflected in the widespread occurrence of bird-flowered species in the genera to which our southwestern species belong.

It seems likely, therefore, that our bird-flowered species with southern affinities, and their ancestors, may have the oldest continuous history of bird pollination of any plants now living in western North America.

The western bird-pollinated plants with northern floristic affinities are better known and better understood. Here we can infer the general sequence in the formation of bird flowers, in a relative way, from the magnitude of the bird-flowered plant group in the taxonomic hierarchy. The following discussion is confined to this northern element in the assemblage of western American bird flowers.

The only large and widespread genus of hummingbird flowers with a rich development in western North America is Castilleja. The red-bracted, bird-flowered Castillejas are represented by numerous species in a wide variety of habitats throughout the West, and occur also in eastern North America and western South America. The yellow and purple-flowered Castillejas, which are probably not bird-pollinated, seem to be derived. Hummingbird flowers, as far as we can judge, represent the basic condition in this large and widely distributed genus. The pattern in Castilleja suggests that it is one of the oldest, if not the oldest extant, group of bird flowers with northern floristic affinities in western America.

We have three minor genera composed wholly or predominantly of bird-flowered species in the western flora: Zauschneria, Diplacus, and Galvezia. Each of these minor genera is allied to a large, predominantly bee-pollinated genus which is well-developed in the west. Zauschneria is allied to Epilobium, Diplacus to Mimulus, and Galvezia to Antirrhinum and its relatives.

It is reasonable to suppose that Zauschneria, Diplacus, and Galvezia budded off from their ancestral bee-pollinated stocks, as new phyletic lines fitted for bird pollination, at some stage later than the origin of Castilleja.

Some 17 large or medium-sized genera in the western flora have one or two bird-flowered species each. The latter stand out in each genus as deviant forms among a large mass of species belonging to some other pollination system—usually bee pollination but sometimes Lepidoptera or fly pollination. The genera containing one or two exceptional species of hummingbird flowers are:

Silene	Gilia	Pedicularis
Delphinium	Polemonium	Scrophularia
Ribes	Monardella	Lobelia
Astragalus	Satureja	Brodiaea
Echinocereus	Trichostema	Fritillaria
	Mimulus	Lilium

The pattern described above points to an even more recent origin of hummingbird flowers in these genera.

Of more recent origin still would be those localized races of humming-bird-pollinated plants which belong to species pollinated by other agents elsewhere. Two examples are *Lonicera involucrata ledebourii* and the San Bernardino Mountain race of *Gilia splendens*.

This brings us to a significant but insufficiently understood group of cases. The flowers of some species are visited and pollinated by both hummingbirds and some other class of agents, usually bees. The flowers exhibit some of the characteristics of hummingbird flowers in a rudimentary state of development. This is the case in *Dudleya cymosa minor* (Crassulaceae), in certain forms of *Phacelia minor* (Hydrophyllaceae), and in *Sarcodes sanguinea* (Grant and Grant, 1966, 1967a). Here we may be witnessing a transitional stage in the evolutionary development of hummingbird flowers.

Hummingbirds feed regularly and systematically on the flowers of Arctostaphylos (Ericaceae) in a wide range of habitats from the high mountains to the desert borders (Plate 28). Sprague (1962) reports similar hummingbird visits to Arbutus of the same family. Some bill-tip pollination must result from this feeding activity. The shrubby Ericaceae have produced bird flowers in other areas, as exemplified by Erica in South Africa (Loew, 1904-1905). Why has no parallel development of bird flowers taken place in Arctostaphylos or other woody members of the Ericaceae in western North America?

Arctostaphylos, Arbutus, and similar woody Ericaceae in the western flora are long-lived plants with a slow turnover of generations. Therefore they require a long period of chronological time to respond to natural selection. In the same length of absolute time, soft-wood subshrubs and many perennial herbs, with their shorter life cycles, will pass through many more generations of selection than will the Manzanitas and their relatives.

The historical duration of the interactions between hummingbirds and western American floristic elements with northern affinities has been long enough to yield hummingbird flowers among plants belonging to the life-form classes of perennial herbs and soft-wood shrubs or subshrubs. Perhaps this same period of time has not been sufficient for the evolution of parallel adaptations to hummingbird pollination by the woody Ericaceae with their long life cycles. It is probably significant that the one member of the Ericaceae in western North America which shows some rudimentary characteristics of a hummingbird flower, and may be transitional, is *Sarcodes sanguinea*, a perennial herb.

All the evidence points to the conclusion that the origin of hummingbird flowers has been a continuing process in western North America, and one which may still be going on.

9 Effects of Hummingbird Migration on Plant Speciation

In certain parts of the mountainous West, the hummingbird-pollinated plant species occur in two contrasting conditions as regards their spatial and ecological relations with one another. The bird-flowered plant species which grow in the lowlands and foothills, and bloom in winter and spring, tend to occur singly or in pairs. By contrast, the bird-pollinated plants which grow in the neighboring high mountains and bloom in summer often occur in flocks of several co-existing species (Grant and Grant, 1967b).

A correlated condition exists among the hummingbirds themselves in the same areas. Many high mountain regions of western North America are summer collecting grounds for hummingbirds which have been widely dispersed over areas of lower elevation in late winter and spring.

We shall attempt to show in this chapter (which is a revised version of an earlier journal article on the same problem—Grant and Grant, 1967c) that the observed correlation is more than a mere coincidence. It seems probable that the seasonal migratory movements of hummingbirds in some types of mountainous country have brought about the characteristic patterns of plant speciation found in the same regions.

Among regions in western North America which display the pattern of plant speciation mentioned above, California is the most familiar to us and the best documented. Therefore the thesis will be developed here in terms of the situation prevailing throughout most of California. We will also consider briefly some other situations in other areas.

Six species of hummingbirds breed within California, and a seventh, the Rufous, migrates through California to its breeding range in the Pacific Northwest. All of these species are migratory, with the exception of the resident Anna hummingbird and a non-migratory island race of the Allen hummingbird (see Chapter 2).

As a general rule, these birds migrate through the Colorado Desert and along the Pacific slope in late winter and spring to their various breeding regions. Then, in the post-breeding season in summer, many of them ascend to the higher mountains of California to feed on the abundant flowers blooming at higher elevations at this time, and later migrate southward (Bent, 1940; Grinnell and Miller, 1944). Of course the details are more complex in certain cases.

The Rufous hummingbird, for example, is an early spring migrant through the lowlands and foothills of the Pacific slope, whereas in the summer its southward migration (in California) is mainly along the higher mountain ranges. Similarly, the Calliope hummingbird migrates northward through California along the foothills of the Pacific slope, breeds in the higher mountains, and then follows the high mountains southward again. The Allen hummingbird migrates northward along the Pacific coast to its breeding range along the coast line, and then in the summer some individuals of this species move to the interior and follow the higher mountains southward. The Black-chinned hummingbird ranges through the deserts and foothills in winter and spring, but in summer, after the breeding season, may be found in the higher mountains, especially those of southern California. Although the Anna hummingbird is resident in California, many individual birds ascend to higher elevations in the mountains in the post-breeding season.

SEASONAL CYCLES IN DISPERSION OF HUMMINGBIRDS

The distribution of hummingbirds in California may be seen in two contrasting phases. In late winter and spring the hummingbirds are relatively well-dispersed over large areas, whereas in summer many hummingbirds become centralized in the higher mountains.

Two sets of factors contribute to these seasonal differences in the density of the hummingbird populations. The first set of factors consists of several aspects of the reproductive behavior and potential of the birds themselves— their breeding ranges, nesting habitat preferences, territorial activities, and fecundity. The second important factor is the physical features of their terrain.

The breeding ranges, nesting habitat preferences, and territorial activities bring about a maximum dispersion of the California hummingbirds in the breeding season. The species are frequently segregated into different ecological zones during the nesting season. Thus the Costa hummingbird typically nests in the arid foothills, the Allen in the humid coastal belt, the Calliope in the high mountains, and so on.

Some pairs of hummingbird species have overlapping breeding ranges. The Anna hummingbird, for example, overlaps with the Allen in some parts of its breeding range, and with the Costa in others. In these cases, however, each species displays different habitat preferences during the breeding season (Pitelka, 1951a, 1951b).

Furthermore, during the breeding season habitat separation between the two sexes of a single species occurs in several species. This is the case in the Allen, Black-chinned, Calliope, and possibly the Broad-tailed hummingbird (Pitelka, 1951b).

On the other hand, during migration territorial activity subsides, and male and female hummingbirds of the same species may then concentrate and feed together in neutral tracts (Pitelka, 1942). Furthermore, several species of hummingbirds are frequently seen feeding together on such common feeding grounds in the spring and summer migrations. To cite a typical example, we have seen Calliope, Rufous, and Allen hummingbirds feeding together in the high Sierra Nevada in summer. The birds thus temporarily form dense populations in sympatric combinations during migration.

In the post-breeding season, moreover, the total number of humming-birds is undoubtedly greater than in the preceding spring and winter, owing to the addition of many new immature individuals. This too must contribute to the increased density of the hummingbird populations in summer.

Turning now to the second factor, there is a great difference in the extent of regions of low to intermediate elevation and the amount of high mountain area in California. The spring breeding ranges of the hummingbirds cover an extensive portion of the state. The northward migration of many hummingbirds takes place over the broad Colorado Desert and wide coastal plain of southern California, and up the foothills and valleys of the Pacific slope (Map 5). But the regions of high mountains, supporting summer-blooming flowers on which southward migrating hummingbirds feed, are, while extensive enough, much less so than the lowland areas (Map 5).

The high mountains are therefore the collecting grounds for birds which have been spread out over much larger areas of lower elevation during the northward migration and the spring breeding season. The physical situation has the effect of concentrating the hummingbirds into dense populations in the summer season.

SEASONAL DISTRIBUTION OF
BIRD-FLOWERED PLANT SPECIES

In an area of varied climate and topography such as California, favorable growth conditions for plants—bird-pollinated as well as others—occur at different seasons in different regions. Plant species which bloom in late winter and early spring predominate in the low deserts, spring blooming species in the foothills and intermediate elevations, and summer blooming species in the high mountains.

We can group the hummingbird flowers of southern California into these three seasonal classes—late winter, spring, and summer—and cite examples of each. In late winter we find *Fouquieria splendens* and *Beloperone californica* in bloom on the Colorado Desert.

In early spring various bird-pollinated species are in bloom on the desert borders or in the low coastal hills:

Desert Borders	*Coastal Hills*
Ipomopsis tenuifolia	Diplacus puniceus
Astragalus coccineus	Salvia spathacea
Penstemon centranthifolius	Pedicularis densiflora

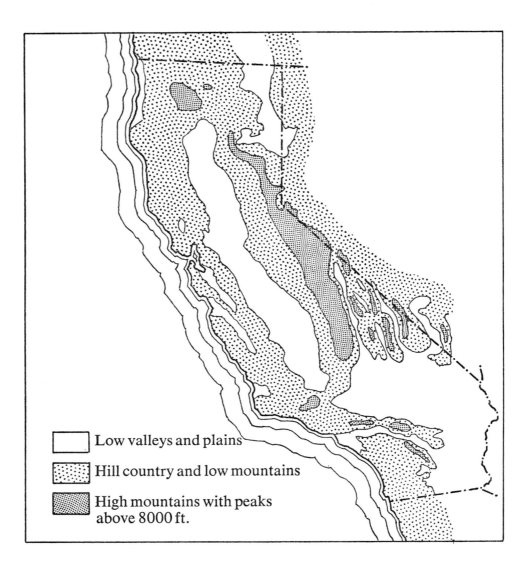

MAP 5. DISTRIBUTION AND EXTENT OF DIFFERENT ELEVATIONAL ZONES IN CALIFORNIA

(From Grant and Grant, 1967c.)

In late spring some other bird-pollinated species come into bloom on the south-facing washes and foothills of the San Gabriel Mountains and neighboring mountain ranges (see Map 4):

Delphinium cardinale
Silene laciniata
Trichostema lanatum

Summer-blooming species occur in the higher mountains of southern California as expected. The following species of hummingbird flowers are in bloom from mid summer into October at higher elevations in the San Gabriel Mountains (Map 4):

Mimulus cardinalis   Penstemon bridgesii
Aquilegia formosa   Zauschneria californica latifolia
Keckia ternata

The seasonal distribution of blooming periods, illustrated here by examples from southern California, can be shown to be general throughout the state (Grant and Grant, 1967c).

ALLOPATRIC VERSUS SYMPATRIC DISTRIBUTIONS OF BIRD-POLLINATED SPECIES IN CALIFORNIA

When we compare different local areas in California, with respect to the numbers of bird-pollinated plant species which grow together, we find a pronounced difference between different seasonal classes of flowers, as mentioned in the introduction to this chapter.

The winter and spring-blooming species of hummingbird flowers are often allopatric, occurring singly in a given territory. Or they frequently occur in a sympatric combination consisting of one dominant and one secondary species.

A typical example of the latter situation among winter bloomers is provided by *Fouquieria splendens* and *Beloperone californica* on the Colorado Desert, where the former is widespread on the desert slopes and mesas, while the latter is restricted to water courses. Again, among spring bloomers in the foothills, *Penstemon centranthifolius* may occur by itself in some localities, or with some species of Castilleja in others.

71

In striking contrast is the situation in the high mountains of California in mid and late summer where half a dozen bird-pollinated species are often found growing very close together and blooming at the same time. In the upper basin of Big Pine Creek in the Sierra Nevada, for example, one can observe the following species of hummingbird-pollinated plants growing either side by side or in adjacent habitats:

Aquilegia formosa — Penstemon newberryi
Ipomopsis aggregata — Castilleja miniata
Penstemon bridgesii — Castilleja breweri

The foregoing general statements are supported by observations in 22 study areas scattered throughout southern and central California and representing a wide range of topographic and climatic conditions (Grant and Grant, 1967c). Each study area covered an area of natural vegetation about one mile in diameter. In each local area we tabulated the bird-pollinated plant species. We then grouped the study areas by elevational zone and season of blooming. The results are summarized in Table 5.

TABLE 5.   NUMBER OF SPECIES OF HUMMINGBIRD-POLLINATED PLANTS GROWING IN EACH OF 22 LOCAL AREAS ABOUT ONE MILE IN DIAMETER SCATTERED THROUGHOUT CALIFORNIA

(From data of Grant and Grant, 1967c.)

Elevational Position of Local Area	Number of Local Areas Studied	Number of Bird-Pollinated Species Per Local Area: Range	Mean
Lowlands, late winter blooming season	6	1-2	1.5
Foothills, spring blooming season	12	1-2	1.3
High mountains, summer blooming season	4	5-6	5.8

The table shows that one or two species of hummingbird flowers per local area is the norm in the lowland and foothill zones, and among winter

72

and spring bloomers. The summer-blooming bird flowers of the high mountains, by contrast, frequently occur in sympatric flocks of five or six species.

EXPLANATION OF THE OBSERVED PATTERNS
IN CALIFORNIA

The number of hummingbird-pollinated plant species per local area in California has been shown to be correlated with physiography, elevation, season of blooming, and seasonal condition of hummingbirds. No doubt additional correlations could be discovered between number of species of bird flowers and other features such as life form which are related to some of the aforementioned variable factors.

We suggest that the correlation between the number of sympatric bird-pollinated plant species and the seasonal state of dispersion of the birds which pollinate them is especially significant and in a class by itself. The hypothesis is proposed that this particular correlation stems from a cause-effect relationship between the density of hummingbirds in a territory and the opportunities for establishment and reproduction of bird-pollinated plant species (Grant and Grant, 1967c).

The information needed in order to distinguish the postulated causal correlation from the other coincidental correlations is not available at present and will be difficult to get. We are dealing here with several simultaneous variables which elude control. It can be said, however, that the secular ecological factors mentioned above do not appear to provide a sufficient explanation in themselves for the observed differences in number of sympatric bird-pollinated species.

If it is argued that the desert lowlands have simpler plant communities than the uplands, then why do we find fairly numerous sympatric species of bee flowers, but not of bird flowers, in the low-lying deserts? Or if it is argued that the high mountain areas provide more ecological niches suitable for bird-pollinated species than do the lowlands, then why don't we find more species of hummingbird flowers per local area in foothill regions than in the lowlands?

There is, finally, no obvious direct connection between the secular ecological factors and the number of sympatric species of bird-pollinated plants, whereas a close connection must exist between the density of pollinating hummingbirds and the conditions of pollination of bird flowers, which suggests that the latter relationship is primary.

All the complexities of the interactions between hummingbirds and plants are involved in the hypothesis advanced here to explain the occurrence of two contrasting distribution patterns among bird-pollinated species in California. There are the interactions between hummingbird individuals; between hummingbird species and plant species; and between different bird-pollinated plant species.

In the spring breeding season, the territorial and nesting activities of the hummingbirds produce interactions between individual birds of the same species and again between hummingbird species which result in a dispersion of individual birds. This phase can be regarded as one in which the selective pressure exerted by the hummingbirds on the regional flora has led to an increase in individuals of single hummingbird-pollinated plant species. But in the high mountains where the hummingbirds congregate in the post-breeding season the selective pressure which they exert is associated with a marked increase in the number of hummingbird-pollinated plant species per local territory.

As regards the plants, the shift from the equilibrium condition of late winter and spring to the very different equilibrium condition of summer probably involves a series of competitive interactions, both between birds and between plants.

The presence of numerous hummingbirds presumably leads to competition between birds for floral food. The competition may then set up a selective pressure in the regional flora toward the evolution of additional bird-flowered species. The food-seeking visits of hummingbirds are by no means restricted to the flowers of those plant species which are specially adapted for hummingbird pollination, as we have seen in Chapter 8. It may be easier for a plant species to become adapted to hummingbird pollination when it grows and blooms with other plant species which do attract numerous hummingbirds, because in such a situation it is more likely to receive recurring visits by food-seeking birds. The dense congregation of hummingbird species in the high mountains of California in summer thus provides a niche for the budding off of new peripheral hummingbird-pollinated races and species of plants, some of which are narrow endemics in these regions.

It is useful to visualize the migratory movements of the hummingbirds in California as forming an annual cycle with a dispersed phase and a concentrated phase. The two phases are separated in space as well as in time. The concentrated phase of the cycle, repeated year after year, must be in effect a sustained selective pressure favoring the formation and establishment of new races and species of bird-pollinated plants. If this view is correct, we have an interesting situation in which a cyclical movement of hummingbirds exerts a boosting effect on plant speciation with particular force in one phase of the cycle.

THE SITUATION IN OTHER WESTERN REGIONS

The effect of hummingbird migration on plant speciation postulated here for the main part of California depends on the populations and species of hummingbirds undergoing certain seasonal cycles in dispersion. These cycles are determined in a large measure, as we have seen, by the physiographic features in the same parts of California. In attempting to determine how far we can generalize from the California situation to other regions of the American West, therefore, we must first specify the California conditions which are believed to be critical, and then look for similar or different conditions elsewhere.

The important topographic characteristics of California, or at least of the main part of the state, are threefold. First, there is a large area of lowlands and foothills, as previously noted. Second, there are high mountain areas which are small relative to the low-lying areas. Third, the high mountain areas are quite extensive in their own right. The crest of the Sierra Nevada runs for 350 miles, and a series of high ranges in southern California continues the mountain chain (Map 5).

A small isolated mountain peak surrounded by extensive areas of lower elevation presents a quite different situation. The small high areas may not be able to provide sufficient summer forage for the huge population of hummingbirds which inhabited the surrounding country in spring. Then the birds, or most of them, will be forced to go elsewhere in summer. Consequently sympatric flocks of bird flowers may not necessarily develop in this situation.

Mt. Lassen in northeastern California (Map 4) is an isolated peak with a highland area which is small in absolute size and also small relative to the neighboring area of lower elevation. It is significant that we do not find numerous sympatric species of bird flowers blooming in summer on the upper slopes of Mt. Lassen, as we do at comparable elevations in the Sierra Nevada. Instead, in the four study areas examined, the bird-flowered species occur in sympatric pairs or singly. The summer pattern of hummingbird flowers on Mt. Lassen is like the spring pattern in other parts of California.

Small isolated high areas similar in a general way to Mt. Lassen are common in the Cascade Mountains and the Great Basin. It will be the task of future field studies to determine whether the Mt. Lassen pattern of distribution of hummingbird flowers is also found in these regions.

The converse topographic situation is that of extensive highland areas and relatively small adjacent lower areas, as in the Colorado Rockies. Such highland areas cannot act as a summer collecting ground for the hummingbirds which spent the spring in the neighboring foothills. Consequently the breeding species of hummingbirds alone will probably not be able to cause the development of a specially rich summer flora of bird flowers.

The same highland areas are also used as summer migration routes by other hummingbird species which breed farther north, particularly the Rufous. What effect these additional hummingbirds have on the summer-blooming plant populations is not known. The situation in the Rocky Mountains needs further study.

In Arizona and Utah we find a topographic set-up more nearly comparable to that in California. Extensive deserts and foothills surround smaller but ample areas of high mountains in parts of these states. In southern Arizona at least the floristic pattern is also similar to that in California, the number of sympatric species of bird flowers being much greater at high than at low elevations.

10 *Common Red Coloration in Hummingbird Flowers*

As we have seen in the preceding chapters, a common floral color—red—is prevalent among the western American hummingbird flowers. What are the possible explanations for convergence in flower color among these plant species?

Two theories have been proposed to explain the association of red floral colors with adaptations to hummingbird pollination. The prevailing view historically is that hummingbirds have a color preference for red. More recently the hypothesis has been suggested that a common floral color is selectively advantageous to both hummingbirds and hummingbird flowers in a region inhabited by migratory hummingbirds (Grant, 1966).

COLOR PREFERENCES OF HUMMINGBIRDS

The importance of hummingbirds as pollinators was first hypothesized by Delpino and first observed by Fritz Müller in 1870, in Santa Catharina, Brazil (Schimper, 1903). In 1903 Schimper wrote (p. 122),

Undoubtedly these brilliantly coloured pollinators show a preference for red, especially for fiery red colours; in regions where hummingbirds abound, for instance the Antilles, I have rarely seen a woody plant resplendent in the sun with the beauty of its red flowers without also being able to detect, with a little patience, humming-birds on it… This preference for red does not, however, exclude visits to flowers that are differently coloured; for the flowers of the species of Marcgravia that I know are of a dull brownish colour.

Later workers observing the North American hummingbirds frequenting red flowers similarly concluded that hummingbirds preferred red flowers. This widespread assumption was questioned by several workers, and other reasons for the prevalence of red floral colors in hummingbird flowers were put forward. For example, Graenicher (1910) observed that the Ruby-throated hummingbird in Wisconsin completely neglected blue flowers, but extensively visited red flowers, and suggested that *Archilochus colubris* was color blind to the blue end of the spectrum. Pickens (1930) considered the conspicuousness of red and orange flowers against a green background as the major factor in attracting hummingbirds to these brightly colored flowers in the eastern United States. Porsch (1931) suggested that red floral colors were more clearly visible than other colors in the early morning and late evening hours when flower-feeding birds were most active.

Early studies on bird vision (summarized by Kühn, 1929) suggested that diurnal birds are more sensitive to red than is man, about equally sensitive to yellow, and less sensitive to blue and violet. McCabe (1961) concluded that although birds distinguish between colors, the spectrum range is not clearly defined, and that diurnal birds are more stimulated by red-yellow bands, while nocturnal birds are more sensitive to those in the blue-green range. Experimental investigation of the color sensitivity of hummingbirds is much to be desired.

A number of experimental studies on the color preferences of North American and Mexican hummingbird species have been carried out, and these do not support the prevailing view that these birds prefer red to other colors.

In 1913 Sherman conducted an experiment "... to test a supposedly erroneous theory which had been published to the effect that Humming-birds show a preference for red flowers." The Ruby-throated hummingbirds tested in an Iowa garden visited vials of sugar water concealed in artificial flowers without regard to the flower color, and visited bottles of colorless sugar water as well (Sherman, 1913).

Bené (1941) investigated color preference in the Black-chinned humming-bird in Arizona, and found no innate preference for red, but concluded that color preference could be conditioned through training.

Wagner (1946) used Mexican species of hummingbirds as subjects in a similar experiment. Feeding flasks covered with paper of different colors were hung out in a place frequented by many hummingbirds; measurements of the amounts of syrup taken from each flask were made at 30 to 60 minute intervals to determine the frequency of hummingbird visits to the different flasks.

Wagner reported (1946):

I found that the flask most frequently visited was always of the same color as the flower most visited at that particular season. It was purple in July when the purple *Penstemon campanulata* was in full bloom and was visited by hummingbirds almost to the exclusion of other flowers. In October, when the dark blue *Salvia mexicana* was in bloom, the liquid in the dark blue flask was almost entirely used up, while the purple flasks that had been favored in July were frequently left full and untouched. Further evidence was the observation that, at a given place and time, each species of hummingbird would choose the flask of the color that corresponded with the color of its preferred flower then in bloom. Thus, in season, the Mexican Violet-ear showed a preference for red flasks, in accordance with its preference for the red flowers of *Salvia cardinalis*; while the White-ear, which visited the blue flowers of *S. mexicana* almost exclusively, patronized the blue flasks almost exclusively.

Lyerly, Riess, and Ross (1950) attempted a more critical experiment to test color preference in a female Mexican violet-eared hummingbird (*Colibri thalassinus thalassinus*) living in a display cage at the New York Zoological Park. Prior to the experiment, small glass food bottles were placed in the bird's cage twice a day. The morning food consisted of a mixture of honey, milk, beef extract, and a commercial vitamin and protein mixture, and the evening food of a dilute honey solution. During the experiment, the same food mixtures were offered at the same times, but the single colorless glass feeder was replaced by four feeders arranged in a linear series within the bird's cage. Each feeder was coated with nail polish of a different color—yellow, green, blue, and red.

The four colors were changed in position at 24-hour intervals, each color appearing in each position once during a four-day period. The experiment was conducted over a 16-day period, and was so designed that statistical analysis of the three variables—color, position of the feeder, and time of day—was possible.

These workers found that the bird exhibited a decided preference for the feeder positioned nearest its favorite perch, no matter what color it was. This preference was more pronounced in the morning than in the evening

79

periods. There was no evidence of a decided preference for any color, but yellow was chosen less frequently (to a statistically significant degree) than were the other colors. No statistically significant preference differences were exhibited by this bird between the red, blue, and green-colored feeders.

In an experiment using a male Anna hummingbird in a California garden, visits to red, yellow, blue, and green feeders were random, as shown by a chi-square test (Grant, 1966). As in the experiment of Lyerly, Riess, and Ross (1950) this hummingbird showed a decided preference for the end-positioned feeders of a row of four, 79 per cent of its visits being to these. The feeding visits of this bird in the early morning before sunrise gave no indication that one color was frequented more than another in dim light.

The experimental results of Ruschi (1953) using Brazilian hummingbird species are not in agreement with those utilizing North American and Mexican species. This worker reported that all 23 species tested decidedly preferred red over seven other colors in feeding tests. Ruschi does not record data concerning the numbers of visits to the different colors for any of the species.

Additional experimental evidence is needed on color preference of hummingbird species. Critical tests are difficult to design, and past experience and conditioning of individual birds is hard to assess. As pointed out by Lyerly, Riess, and Ross (1950) the assumption that hummingbirds prefer red seems to be based on the high incidence of red floral colors among bird flowers, and experimental evidence fails to show such a preference.

LEARNING IN HUMMINGBIRDS

Observations on the feeding behavior of young Black-chinned hummingbirds indicate that the birds learn by trial and error which particular flowers provide a source of nectar (Bené, 1945).

Adult Black-chinned hummingbirds established an association between particular sites and specific food sources in a single visit to the food source (Bené, 1945). These birds quickly perceived different concentrations of sugar water in feeders, and when these were placed at different heights, quickly learned the position of the feeder containing the strongest syrup (Bené, 1945).

This behavior suggests that hummingbirds, which discriminate between colors, can readily establish an association between colors and food source. Such an association between red colors and food sources would be formed early in the life of hummingbirds which nest in a region where abundant and available nectar is most often found in red flowers.

The many cited examples of hummingbirds curiously investigating various red objects other than flowers is explained by such an association of red and food, as has been pointed out by several earlier workers (for example, Woods, 1927; Jaeger, 1950).

Woods (1927) noted in California:

Hummingbirds are attracted to flowers initially by their coloring. A bunch of carrots will sometimes arouse the interest of a Hummingbird, and I have seen one probing clusters of bright orange Crataegus berries. But when once established in a locality they will habitually pass over some of the showiest flowers as unsuited to their uses, seeking out others, perhaps much less conspicuous, which minister to their needs.

AN EXPLANATION OF COMMON RED COLORATION

In western North America most of the hummingbird species are migratory, in contrast to the typical resident status of the Trochilidae in tropical regions (Ridgway, 1891). As discussed in Chapter 9, these western American species not only migrate large distances in spring and fall, but undergo a seasonal dispersion in western America as well, often ascending to higher elevations in summer in search of new sources of floral food.

Most of the plant species of western North America dependent on hummingbirds for their pollination display a common floral color. Grant (1966), in considering the situation in California, suggested that this common red coloration would be collectively beneficial to these plant species. The display of a common floral color by the group of hummingbird-flowered species serves to pool an advertisement of food sources to the hummingbirds. In this way the flower-feeding visits of the hummingbirds are channeled toward those species dependent on them for pollination rather than being distributed as well among plant species normally pollinated by other agents. Each plant species sharing in the common red coloration would then benefit in sharing in the pollinating visits of hummingbirds.

Hummingbirds, in learning to associate red floral colors with availability of large quantities of nectar, would benefit in quickly finding new floral food sources through recognition of their red color. This ability to find new food sources without going through a trial and error process would be particularly advantageous to migratory hummingbirds which continually enter new feeding territories (Grant, 1966).

The analogy of warning coloration in animals is helpful for an understanding of common red coloration. The well-known phenomenon of Müllerian mimicry, in which a group of often unrelated species advertise their unpalatable or poisonous qualities to predators by a similar conspicuous color pattern, permits the various protected species to pool the losses of individuals involved in the education of predators, and to share the protection gained when the predators learn to associate the warning coloration with distastefulness or poisonousness.

Common red coloration represents a comparable type of adaptive coloration on the part of a diverse assemblage of plant species dependent on a common pollinating agent. It is reasoned that this display of a common floral color is selectively advantageous to both the plants and their bird pollinators in that it facilitates quick recognition of food sources in the whole area of migration.

Several factors may be involved in contributing to the occurrence of red as the common floral color. As discussed in Chapter 8, many of the bird-flowered species of the western American flora are derived from bee-pollinated ancestors. The change-over from bee-pollinated forms to bird-pollinated forms has most often been accompanied by a change in floral colors, commonly from blue to red.

The change from a floral color highly attractive to bees to one which is not attractive to these insects (Kühn, 1929) may serve to augment the development of a flower form adapted to bird-pollination from a bee-pollinated ancestor.

The occurrence of red floral colors, however, is not restricted to those species derived from bee-pollinated ancestors. For example, Silene, a genus in which Lepidoptera pollination is prevalent, contains two western American hummingbird-pollinated species (*Silene laciniata* and *S. californica*), both of which bear bright red flowers.

The color perception of hummingbirds may also be involved, for red may be a color highly visible to these birds as it is to man. But this is not sufficient in itself to explain common red coloration, since, as already pointed out, hummingbirds will readily visit flowers or feeders of many other colors.

In conclusion, the evolution of common red coloration in western North American hummingbird flowers has probably been affected by several factors. The primary development has been toward convergence in flower color. The occurrence of red as this common color has probably been influenced by the sense perception both of hummingbirds and competitor flower-feeding bees.

PRELIMINARY TESTS OF THE PROPOSED EXPLANATION

Red floral colors do occur with a high frequency among bird-pollinated species the world over. Several authors, however, have emphasized that red can not be considered the typical color of bird flowers, for many other floral colors are associated with bird pollination.

Porsch (1931) pointed out the relative infrequency of bright red flowers among the bird-pollinated members of the flora of Australia, and the exceptional occurrence of red colors among the bird-pollinated Lobelioids of the Hawaiian Islands. Yellow occurs with greater frequency than red among bird-flowered species occurring in the boundary area between Argentina and Bolivia (Melin, 1935, referring to data of Fries).

Many of the American Bromeliads are bird flowers. Floral colors of these species include yellow, white, blue, green, orange, and brown, bright red being exceptional as a floral color but often appearing on bracts or foliage (Porsch, 1931). In the genus Agave, containing many bird-pollinated forms, yellow and green are the most frequent flower colors, and red is rare (Porsch, 1931).

Thus an array of floral colors is exhibited among bird-pollinated species as a whole.

A similar diversity in floral colors exhibited by hummingbird-pollinated forms would be expected to occur in the American tropics, according to the hypothesis of common red coloration. As mentioned earlier, tropical hummingbird species are resident, spending the entire year in the same area, with only local seasonal movements in search of food (Ridgway, 1891). In this particular situation, the ability to quickly recognize new floral food sources would not have the premium value postulated for the migratory and nomadic north temperate species.

The advantage in tropical regions, where hummingbird species display great diversity in bill size and shape, feeding habits, and ecological habitat preferences, may be in differentiation among sympatric hummingbird flowers.

The tropical American hummingbird flowers are little known, and the incidence of different floral colors among them is a problem requiring much more information based on extensive field studies. Nevertheless, some scattered evidence does point to greater diversity in floral colors among these species.

Floral colors of some tropical Mexican plant species observed by the authors to be visited by hummingbirds (Grant, 1966) include pale pink (*Bomarea acutifolia*), deep magenta (*Maurandya erubescens*, *Penstemon* sp.), bright orange (*Heliconia latispatha*, Loranthaceae), yellow (Bignoniaceae), red and yellow (*Dudleya* sp.), blue perianth and red bracts (Bromeliaceae), bright red (*Hamelia sp.*, *Bouvardia ternifolia*, and *Salvia sessei*).

In South America species of Cantua (Polemoniaceae) visited by hummingbirds display an array of floral colors ranging through red, orange, and magenta to white (Grant and Grant, 1965).

The floral form of the blue-flowered *Stachytarpheta bicolor* (Verbenaceae) strongly suggests that this species is adapted to hummingbird pollination, as pointed out by Dr. S. Vogel (pers. comm.). Skutch (1958, 1967) notes that a number of hummingbird species and especially the Violetheaded hummingbird (*Klais guimeti*) are attracted to the flowers of Stachytarpheta in Central America.

In considering the overall situation in Central America Skutch writes (pers. comm.):

My impression—without having made a statistical analysis—is that hummingbird flowers here in Central America are predominantly red, but not to the same extent as in California; other colors appear to be of relatively more frequent occurrence. Here in the tropics many hummingbirds are resident in the same locality at all seasons, so that they can become well acquainted with the best sources of nectar whatever the floral color may be. On the other hand, hummingbirds as a family tend to wander more widely than do many other families of birds that breed in the humid tropics, so that the possession of a common "hummingbird color" might be of some advantage to the flowers.

If, as postulated, the migratory and nomadic habits of the western American hummingbirds and the occurrence of a common floral color among the hummingbird flowers of this region are related, then a similar situation might be expected in other regions of few and migratory hummingbird species. A single summer-resident hummingbird species (*Archilochus colubris*) occurs in the eastern United States. Red coloration apparently prevails among the hummingbird flowers of this region; according to Pickens (1930) these are either red or orange.

11 *Reciprocal Evolution of Hummingbirds and Plants*

The question before us in this final chapter is: how does a co-adapted system of hummingbirds and flowers develop, in the course of evolution, to an advanced state of mutual specialization?

It can be said that hummingbirds are a selective factor leading to floral specialization. But this is an incomplete answer. It does not account for the other side of the co-adapted system. Nor does the complementary incomplete answer—namely that hummingbirds have gone through many generations of selection for and adaptation to flower feeding—give us a satisfactory solution to the problem. Nor do the two partial answers add up simply to one complete explanation. We have somehow to explain the evolutionary development of the two-sided plant-animal relationship as a whole.

Our attempt to answer this question here necessarily takes the form of a hypothesis. Some facets of the hypothesis are supported by evidence already reviewed in previous chapters. Other facets may prove difficult to verify because we are dealing with a long-term historical process. Still others have not as yet been tested but could be in future studies.

INTERACTIONS BETWEEN PRE-ADAPTED BIRDS AND FLOWERS

Our starting point has to be a stage just prior to the existence of a bird pollination system in the American hemisphere. The ecosystem then included an array of flowers adapted for pollination by agents other than birds, and an array of birds adapted for obtaining food from sources other than flowers. It can be safely assumed from observable events in the living world today that the non-flower birds occasionally visited the non-bird flowers in search of small insects or nectar or both.

Only a fraction of these illegitimate visits were to have evolutionary consequences. Our story is concerned with this small but evolutionarily significant fraction of the illegitimate visitations.

Among the bird species making exploratory floral visits, a few already possessed structural and behavioral traits enabling them to successfully exploit this new source of food. They were, as we can say in retrospect, "pre-adapted" for feeding on flowers. Some of these pre-adapted birds, moreover, may well have been under pressure of competition in their old food niche to invade the new food niche.

In the event, some primitive hummingbird or swift-like bird—tropical, insectivorous, and agile on the wing—was to become the ancestor of the nectar-feeding hummingbirds. It is tempting to speculate that the insectivorous diet of the swifts and primitive hummingbirds may have driven the ancestors of the nectariferous hummingbirds to seek insects in flowers; that their fine bills were fitted for taking such prey along with nectar out of the flowers; and that their habit of catching insects in flight enabled them to hover briefly at least in front of the flowers while feeding.

The gradual transition to a predominantly nectar diet is not difficult to account for. Nectar is a desirable type of food for many non-nectariferous birds. Linnets, orioles, and other birds eagerly feed on sugar-water when it is offered in bird feeders, and sometimes seek nectar from flowers in nature. The demand for nectar is evidently more widespread among bird species than the ability to get it successfully. The pre-adapted ancestors of the modern hummingbirds which were able to extract nectar from flowers would have had an advantage over competitor species in invading this food niche and in evolving true adaptations for flower feeding.

We turn our attention next to the array of plant species whose flowers were visited by the earliest American flower-feeding birds. Here too it is likely that some were pre-adapted, by virtue of their existing floral structures, for bird pollination, and some were not.

We saw in Chapter 8 that the hummingbird flowers of relatively recent origin in western North America are a non-random sample of the parental western flora. The actual sample of bird flowers is biased strongly in favor of sympetalous flowers related to and probably derived from bee-pollinated species. Other flower forms and other ancestral pollination systems are represented rarely or not at all in this sample. The observed biases are explicable in part at least as a reflection of differences between plant species in the potentiality for evolving functional hummingbird flowers. It is logical to suppose that similar differences between plant species in their pre-adaptations for bird pollination existed in tropical American floras of an earlier age.

Hummingbird pollination offers some advantages to plants in its own right. Hummingbirds are active feeders and make numerous floral visitations; they carry pollen readily on their heads and bills and bring about much flower pollination. The intrinsic advantages may be enhanced by external factors in certain times and places. If a different and older pollination system—some form of insect pollination—has reached a high level of development in the biome, so that plants are in competition for members of that class of pollinators, various pre-adapted plant species would be favored in their seed production by evolving attractive and pollination mechanisms adapted to hummingbirds.

DEVELOPMENT OF INCIPIENT FUNCTIONAL SPECIALIZATIONS

A plant population adapted for pollination by some class of insects and pre-adapted to hummingbirds, and receiving a greater number of effective pollinating visits by the latter than by the former, will be impelled by natural selection to develop floral mechanisms which function ever more efficiently with the bird visitors. Similarly, the primitive flower-visiting hummingbird will be impelled by natural selection to improve its structural and behavioral adaptations for feeding on flowers.

The results of these processes can be seen in functional specializations for bird pollination and for flower feeding, respectively, which are imperfect in a transitional stage.

Floral mechanisms which are transitional to full-fledged hummingbird flowers are known in several instances, as noted in Chapter 8. Among the examples are *Dudleya cymosa minor*, *Dudleya lanceolata* (Crassulaceae), and *Sarcodes sanguinea* (Ericaceae). Such species have attractive and pollination mechanisms which diverge from the syndrome of bee- or other insect-pollinated flowers, and approach but do not fully attain the typical condition of hummingbird flowers.

Little is known about transitional specializations for flower feeding in the birds. Among hummingbirds these could be looked for in the short-billed tropical species; and examples might be found also in other tropical American families of flower-visiting birds such as the Coerebidae.

DEVELOPMENT OF EXCLUSIVENESS

The system composed of trumpet-shaped bird flowers and long-billed hummingbirds, as found widely throughout temperate North America and elsewhere, has reached an advanced stage of mutual specialization.

The trumpet-shaped hummingbird flowers and the long-billed hummingbirds comprising this system are clearly co-adapted. However, it is obvious that short-tubed flowers and short-billed hummingbirds could be just as well co-adapted. Why then does the co-adapted system of birds and flowers which consists of members with elongated tubes and bills take the particular form that it does?

This system represents a stage in the development of exclusive mutual relationships between flowers and birds. North American bird flowers like the red Penstemons and Aquilegias, with their long slender tubes and flexible flower stalks, not only yield nectar to their hummingbird visitors, but also deny it to most other kinds of nectar-seeking animals.

Of course, the exclusiveness is not complete. Hawkmoths can obtain the nectar of these bird flowers by hovering and probing with their long tongues, and the larger bees by biting holes in the base of the floral tube. Pollination is often carried out by pollen-collecting bees (see Macior, 1966). Nevertheless, the exclusiveness is developed to a relatively high degree, in that most potential nectar-seeking insects and other animals are kept out of the nectar chamber.

The exclusive mechanisms of a hummingbird flower have advantages and disadvantages. The advantage lies in conserving the nectar supply for the hummingbirds, thus enhancing the attractive power of the flowers for a class of visitors of proven value in pollination. The corresponding disadvantage is of course the loss of some potential insect pollinators. Since exclusive mechanisms have in fact evolved in hummingbird flowers, it is logical to assume that their selective advantage has outweighed their disadvantage.

There are strong indications, as we attempted to show in Chapter 8 for the western American flora, that many species of hummingbird flowers have evolved from insect-pollinated ancestors. Competition between plant species for flower-visiting bees and other insects is implicated in the evolutionary change-over of certain plants to hummingbird pollination. Adaptation to hummingbirds is a way out of the competition for insect pollinators. And the exclusive floral mechanisms which reserve the nectar for the birds but exclude most insects are likely to bring about more gains than losses in the pollination of those plants which have been relatively unsuccessful in the competition for insects.

Our next problem is to explain the historical beginnings of an exclusive mechanism consisting of long slender floral tubes. The problem comes into focus when we consider the following propositions. Trumpet-shaped flowers have evolved from shorter-tubed ancestors which were presumably fed on by insects as well as birds. Such trumpet flowers now provide nectar for long-billed hummingbirds which have descended from shorter-billed predecessors; and at the same time they exclude many modern nectar-feeding insects such as bees with shorter tongues. Why haven't the bees and other insects kept pace with the birds in the evolutionary development of increasingly long bills? Why, in other words, has the long slender nectar tube been able to succeed as an exclusive mechanism in hummingbird flowers?

It would be very difficult to account for the first step in the evolutionary trend to a longer and more slender shape in hummingbird flowers if the primitive insects and birds feeding on the ancestral flowers had bills of equal length. This difficulty disappears if the primitive hummingbirds had a longer reach from the first than the bees and other insects feeding on the same flowers. Perhaps the primitive hummingbirds acquired certain bill and tongue proportions, as an adaptation for picking insects out of flowers, which enabled them to probe deeper than most insects. The first step to a slightly longer and more exclusive bird flower would then be possible. And this first step, once taken, would open the way for the progressive evolution of still longer floral tubes fitted for still longer bird bills.

Müller (1883) pointed out that bees use their mouth-parts in constructing cells and suggested that this use may have set definite limits to the elongation of their proboscis. It is possible that this factor may have prevented bees from keeping pace with hummingbirds in the evolution of longer bills fitted for probing into longer-tubed flowers.

SPREAD OF EXCLUSIVENESS

The account just given will explain the elongation of bird bills and floral tubes in a single hummingbird species and in one or more plant species lying within its distribution area. It will also explain the repeated and independent occurrence of the same process in different areas and involving different bird and plant species. But it does not provide a sufficient explanation, in itself, for the whole situation as we find it in western North America.

In western North America there are seven or more species of hummingbirds, belonging to several different genera, all with long slender bills of closely similar proportions. And there are numerous unrelated plant species in such genera as Penstemon, Ipomopsis, Zauschneria, and Aquilegia with long slender floral tubes of a similar and matching length. The co-adapted system extends to numerous species of plants and birds living in the same area.

The simplest way of explaining this situation is to postulate that a given exclusive bill-tube system developed once in some original bird species and plant species under the guidance of the factors outlined in the preceding section. This original bird-flower combination was successful. Its successfulness brought a new factor into the picture.

Other hummingbird species in search of food took to visiting the original flower species regularly. Consequently they came under natural selection for convergence toward the original bird species in bill length. Conversely, additional plant species took advantage of the opportunities created by the existence of additional flower-visiting hummingbirds, and they too evolved convergent floral characters under the influence of selection.

In this way a given exclusive co-adapted system would accumulate more and more plant species and hummingbird species. The system would spread throughout the flora and the hummingbird fauna of a region. Its spread leads to a regional ecosystem consisting, on the one hand, of several hummingbird species able to feed equally well on any native species of hummingbird flower, and on the other hand, of numerous species of bird flowers which can be pollinated successfully by any species of hummingbird.

ON TO NARROWER SPECIALIZATION

The situation just described, with broad standardization and interchangeability of parts between different species of birds and plants, prevails in western North America. There is reason to believe that the situation in some regions at least of the American tropics is very different.

Our first clue comes from an inspection of museum collections of tropical hummingbirds. Here, in marked contrast to the uniformity of bill proportions in the western American species, we observe much interspecific diversity in bill length and shape. There are short, long, and very long bills; straight, slightly curved, and sickle bills; and so on. This diversity suggests specialization among tropical hummingbirds with regard to food plants.

We have very little actual evidence about the feeding habits of tropical hummingbirds, and this is a problem, or group of problems, which richly deserves investigation. A study by Skutch (1952) in Costa Rica does, however, lend support to the idea of some food-plant specialization between tropical species of hummingbirds.

The Scarlet Passion-flower (*Passiflora vitifolia*) grows in tall second-growth woods in Costa Rica. Although this vine climbs up to the canopy of the forest where the main part of the foliage grows in the sun, the large brilliant red flowers are borne on special shoots in the deep shade of the forest near the ground. Skutch (1952) found that although at least 20 species of hummingbirds occurred in the locality studied, only one hummingbird species, the Buff-browed hermit (*Phoethornis superciliosus*), regularly visited and pollinated the flowers.

The Buff-browed hermit lives in the primary forest and second-growth woodland, and stays beneath the forest canopy in the deep shade where the Scarlet Passion-flowers bloom. A regular visitor to the flowers, *Phoethornis superciliosus* effects their pollination while probing for nectar with its long, curved bill.

Although several other hummingbird species in the region have bills long enough (at least 3 cm) to reach the nectar and pollinate the flowers, these are birds of higher elevations which rarely descend to the habitat of the Passiflora flowers. One other hummingbird, Longuemare's hermit (*Phoethornis longuemareus*), also a species of the dark undergrowth, regularly visited the Passiflora flowers. But this bird's bill was too short to reach the nectar in the flowers, and it either sipped nectar from the nectaries on the floral bracts or picked insects, in either case failing to pollinate the flowers.

Skutch (1952) concluded that, as a result of both the bill lengths and the habitat preferences of the hummingbird species, *Passiflora vitifolia* is pollinated chiefly by *Phoethornis superciliosus* in the area studied.

We would expect narrow specializations within the hummingbird pollination system to arise in tropical regions where this system has been able to undergo a long development and reach a mature state. Specialized bird-flower relationships are an expected result of long-continued interspecific competition between birds for floral food, and between plants for bird pollinators.

GENERAL CONCLUSIONS

In this chapter we have attempted to explain the evolutionary development of a specialized pollination system in terms of hummingbirds and their flowers. We see this process of development as a sequence of stages and have treated it as such. It is useful now to review the probable developmental sequence as a whole.

A co-adapted plant-animal system has its inception, its transitional period, its intermediate period of broad specialization, and its advanced stage of narrower specializations. Running through the sequence of stages is a common thread. This thread we will call (for want of a better expression) the density of the pollination system.

The density of a pollination system is measured by the number of plant and animal species participating more or less regularly in that system. Undoubtedly the relative density of pollination systems, thus defined, varies widely from one geographical region to another; and it probably varies over a similar wide range during biotic evolution from one time period to another. The differences in density of a pollination system reflect, in turn, differences in the expected intensity of the evolutionarily significant factor of competition between plants for pollinators and between animals for flower food. (For a discussion of the problem of competition in natural communities and in evolution, we refer the reader to Lack, 1966.)

The hummingbird pollination system is thought to have had its inception in a biotic environment in which the existing insect pollination system(s) had reached a high density. A highly developed bee pollination system was evidently the breeding ground of many new lines of bird-pollinated plants in western North America. We have no way of knowing what type or types of insect pollination systems were the precursors of the more ancient hummingbird flowers in the American tropics.

The hummingbird pollination system in its early stages must have consisted of only a few species of birds and plants, as in the Pacific Northwest today. An intermediate stage of development of bird pollination is correlated with an intermediate density of the system in western North America. In the American tropics where the hummingbird pollination system has existed for the longest time and reached its greatest density, it has also developed to its most advanced stage of specialization.

Hummingbird pollination in some tropical regions may have reached a dense or saturated condition which is conducive to the emergence of a new derivative pollination system. And we may be seeing signs of the emergence of such a new derivative system in the sporadic cases of bat pollination. Bat pollination is clearly an advanced condition in the Angiosperms.

In a number of tropical American plant groups, bat flowers and hummingbird flowers occur together in the same genera or in related genera of the

94

same family. This pattern is found inter alia in Matisia (Bombacaceae), Bauhinia (Caesalpiniaceae), Parkia (Mimosaceae), Calliandra (Mimosaceae), Mucuna (Papilionaceae), probably Marcgravia (Marcgraviaceae), maybe Drymonia (Gesneriaceae), and Agave (Agavaceae); and at a higher taxonomic level in such families as the Melastomataceae, Bombacaceae, Sterculiaceae, Solanaceae, and Bignoniaceae. The systematic distribution of these bat flowers suggests that they are derived from bird flowers. Thus the course of evolution of the hummingbird pollination system has run full cycle.

Bibliography

Abrams, L. 1940-1960. Illustrated Flora of the Pacific States. 4 vols., Stanford University Press, Stanford, Calif.

Anderson, J. P. 1959. Flora of Alaska and Adjacent Parts of Canada. Iowa State University Press, Ames.

Austin, O. L. 1961. Birds of the World. Golden Press, New York.

Beeks, R. M. 1962. Variation and hybridization in southern California populations of Diplacus (Scrophulariaceae). Aliso 5:83-122.

Bené, F. 1941. Experiments on the color preference of black-chinned hummingbirds. Condor 43:237-42.

———— 1945. The role of learning in the feeding behavior of black-chinned hummingbirds. Condor 47:3-22.

———— 1947. The feeding and related behavior of hummingbirds with special reference to the Black-chin, *Archilochus alexandri* (Bourcier and Mulsant). Mem. Boston Soc. Nat. Hist. 9 (3):399-478.

Bent, A. C. 1940. Life histories of North American cuckoos, goatsuckers, hummingbirds, and their allies. U. S. Nat. Mus. Bull. 176. (Dover edition, 1964.)

Darwin, C. 1862. On the Various Contrivances by which Orchids are Fertilized by Insects. John Murray, London.

Davis, J. S. 1956. Natural pollination of California lilies. Thesis, Claremont Graduate School, Claremont, Calif.

Davis, R. J. 1952. Flora of Idaho. W. C. Brown Co., Dubuque, Ida.

Dawson, W. L. 1923. The Birds of California. Vol. 2. South Moulton Co., San Diego, Los Angeles, and San Francisco.

Epling, C., and H. Lewis. 1952. Increase of the adaptive range of the genus Delphinium. Evolution 6:253-67.

97

Faegri, K., and L. van der Pijl. 1966. The Principles of Pollination Ecology. Pergamon Press, Oxford.

Frisch, K. von. 1950. Bees, Their Vision, Chemical Senses, and Language. Cornell University Press, Ithaca, N.Y.

Graenicher, S. 1910. On Humming-bird flowers. Bull. Wisc. Nat. Hist. Soc. 8:183-86.

Grant, K. A. 1966. A hypothesis concerning the prevalence of red coloration in California hummingbird flowers. Amer. Nat. 100:85-98.

Grant, K. A., and V. Grant. 1964. Mechanical isolation of *Salvia apiana* and *Salvia mellifera* (Labiatae). Evolution 18:196-212.

—— and —— 1967b. Records of hummingbird pollination in the western American flora. III. Arizona records. Aliso 6 (3):107-10.

—— and —— 1967c. Effects of hummingbird migration on plant speciation in the California flora. Evolution 21:457-65.

Grant, V. 1950. The protection of the ovules in flowering plants. Evolution 4:179-201.

—— 1952. Isolation and hybridization between *Aquilegia formosa* and *A. pubescens*. Aliso 2:341-60.

—— 1958. The regulation of recombination in plants. Cold Spring Harbor Symp. Quant. Biol. 23:337-63.

—— 1961. The diversity of pollination systems in the Phlox family. Recent Advances in Botany (Toronto) 1:55-60.

Grant, V., and K. A. Grant. 1965. Flower Pollination in the Phlox Family. Columbia University Press, New York and London.

—— and —— 1966. Records of hummingbird pollination in the western American flora. I. Some California plant species. Aliso 6 (2):51-66.

—— and —— 1967a. Records of hummingbird pollination in the western American flora. II. Additional California records. Aliso 6 (3):103-5.

Greenewalt, C. H. 1960. Hummingbirds. (Published for The American Museum of Natural History). Doubleday and Co., Garden City, N.Y.

Grinnell, J., and A. H. Miller. 1944. The distribution of the birds of California. Pac. Coast. Avi., No. 27:1-608.

Harrington, H. D. 1954. Manual of the Plants of Colorado. Sage Books, Denver, Colo.

Henry, J. K. 1915. Flora of Southern British Columbia and Vancouver Island. W. J. Gage and Co., Toronto.

Hitchcock, C. L., A. Cronquist, M. Ownbey, and J. W. Thompson. 1955-64. Vascular Plants of the Pacific Northwest. Vols. 2-5. University of Washington Press, Seattle.

Hultén, E. 1941-1949. Flora of Alaska and Yukon. Lunds Universitets Årsskrift, vols. 37-46. Lund.

Jaeger, E. C. 1940. Desert Wild Flowers. Stanford University Press, Stanford, Calif.

—— 1950. Our Desert Neighbors. Stanford University Press, Stanford, Calif.

Jepson, W. L. 1922-1943. A Flora of California. 3 vols. Associated Students Store, Berkeley, Calif.

——— 1925. A Manual of the Flowering Plants of California. Associated Students Store, Berkeley, Calif.

Kearney, T. H., and R. H. Peebles. 1960. Arizona Flora. Ed. 2, University of California Press, Berkeley.

Kerner, A. 1894-1895. The Natural History of Plants. 2 vols., (English transl.). Blackie, London.

Knoll, F. 1921-1926. Insekten und Blumen. Abhandl. zool.-bot. Gesellsch. Wien, vol. 12.

Knuth, P. 1906-1909. Handbook of Flower Pollination. 3 vols., (English transl.). Oxford University Press, Oxford.

Kugler, H. 1955. Einführung in die Blütenökologie. Gustav Fischer, Stuttgart.

Kühn, A. 1929. Farbenunterscheidungsvermögen der Tiere, p. 720-41. *In* A. Bethe, G. v. Bergmann, G. Embden, u. A. Ellinger [eds.], Handbuch der normalen und pathologischen Physiologie. Vol. 12, Part 1. Springer, Berlin.

Kullenberg, B. 1961. Studies in Ophrys pollination. Zoologiska Bidrag från Uppsala, vol. 34.

Lack, D. 1966. Population Studies of Birds. Clarendon Press, Oxford.

Lasiewski, R., F. R. Galey, and C. Vasquez. 1965. Morphology and physiology of the pectoral muscles of humming-birds. Nature 206:404-5.

Loew, E. 1904-1905. Handbuch der Blütenbiologie. Vol. 3 of Knuth's Handbuch. Wilhelm Engelmann, Leipzig.

Lyerly, S. B., B. F. Riess, and S. Ross. 1950. Color preference in the Mexican Violet-eared hummingbird, *Calibri t. thalassinus* (Swainson). Behaviour 2:237-48.

Macior, L. W. 1966. Foraging behavior of Bombus (Hymenoptera: Apidae) in relation to Aquilegia pollination. Amer. Jour. Bot. 53:302-9.

Mayr, E. 1964. Inferences concerning the Tertiary American bird faunas. Proc. Natl. Acad. Sci. 51:280-88.

McCabe, R. A. 1961. The selection of colored nest boxes by House wrens. Condor 63:322-29.

Melin, D. 1935. Contributions to the study of the theory of selection. II. The problem of ornithophily. Uppsala Universitets Årsskrift 16:3-355.

Merritt, A. J. 1897. Notes on the pollination of some Californian mountain flowers. IV. Erythea 5:15-22.

Miller, A. H. 1951. An analysis of the distribution of the birds of California. Univ. Calif. Publ. Zoo. 50:531-644.

Müller, H. 1881. Alpenblumen, Ihre Befruchtung durch Insekten. Wilhelm Engelmann, Leipzig.

——— 1883. The Fertilisation of Flowers. (English transl.). Macmillan, London.

Munz, P. A., and D. D. Keck. 1959. A California Flora. University of California Press, Berkeley.

Pearson, O. P. 1950. The metabolism of hummingbirds. Condor 52:145-52.

——— 1954. The daily energy requirements of a wild Anna hummingbird. Condor 56:317-22.

Peck, M. E. 1961. A Manual of the Higher Plants of Oregon. Ed. 2, Oregon State University Press, Corvallis.

Pennell, F. W. 1935. The Scrophulariaceae of eastern temperate North America. Monographs, Acad. Nat. Sci. Philadelphia, no. 1.

Percival, M. S. 1965. Floral Biology. Pergamon Press, Oxford.

Phillips, A., J. Marshall, and G. Monson. 1964. The Birds of Arizona. University of Arizona Press, Tucson.

Pickens, A. L. 1930. Favorite colors of hummingbirds. Auk 47:346-52.

Pitelka, F. A. 1942. Territoriality and related problems in North American hummingbirds. Condor 44:189-204.

——— 1951a. Ecologic overlap and interspecific strife in breeding populations of Anna and Allen hummingbirds. Ecology 32:641-61.

——— 1951b. Breeding seasons of hummingbirds near Santa Barbara, California. Condor 53:198-201.

Porsch, O. 1924-1929. Vogelblumenstudien. Jahrb. Wissensch. Bot. 63:553-706; 70:181-277.

——— 1926-1933. Kritische Quellenstudien über Blumenbesuch der Vögel. Biologia Generalis, 2:217-40; 3:171-206; 3:475-548; 5:157-210; 6:133-246.

——— 1931. Grellrot als Vogelblumenfarbe. Biologia Generalis 7:647-74.

Ridgway, R. 1891. The Humming birds. U.S. Nat. Mus. Report, 1890, p. 253-383.

Ruschi, A. 1953. A cor preferida pelo beija-flores. Boletim do Museu de Biologia (Santa Teresa, Brazil), No. 22, p. 1-5.

Schimper, A. F. W. 1903. Plant-geography upon a physiological basis. (English transl.). Clarendon Press, Oxford.

Shaw, R. J. 1962. The biosystematics of Scrophularia in western North America. Aliso 5:147-78.

Sherman, A. R. 1913. Experiments in feeding hummingbirds during seven summers. Wilson Bull. 25:153-66.

Skutch, A. F. 1931. The life history of Rieffer's hummingbird (*Amazilia tzacatl tzacatl*) in Panama and Honduras. Auk 48:481-500.

——— 1951. Life history of Longuemare's hermit hummingbird. Ibis 93:180-95.

——— 1952. Scarlet Passion-flower. Nature Mag. 45:523-25, 550.

——— 1958. Life history of the Violet-headed hummingbird. Wilson Bull. 70:5-19.

Skutch, A. F. 1964. Life histories of Hermit hummingbirds. Auk 81:5-25.

———— 1967. Life histories of Central American highland birds. Publ. Nuttall Ornith. Club, No. 7:1-213.

Sprague, E. F. 1962. Pollination and evolution in Pedicularis (Scrophulariaceae). Aliso 5:181-209.

Straw, R. M. 1956a. Adaptive morphology of the Penstemon flower. Phytomorphology 6:112-19.

———— 1956b. Floral isolation in Penstemon. Amer. Nat. 90:47-53.

———— 1966. A redefinition of Penstemon (Scrophulariaceae). Brittonia 18:80-95.

Vogel, S. 1954. Blütenbiologische Typen als Elemente der Sippengliederung dargestellt anhand der Flora Südafrikas. Gustav Fischer, Jena.

Wagner, H. O. 1945. Notes on the life history of the Mexican violet-ear. Wilson Bull. 57:165-87.

———— 1946. Food and feeding habits of Mexican hummingbirds. Wilson Bull. 58:69-93.

Weymouth, R. D., R. C. Lasiewski, and A. J. Berger. 1964. The tongue apparatus in hummingbirds. Acta anat. 58:252-70.

Wiggins, I. L., and J. H. Thomas. 1962. A Flora of the Alaskan Arctic Slope. University of Toronto Press, Toronto.

Woods, R. S. 1927. The hummingbirds of California. Auk 44:297-318.

Wooton, E. O., and P. C. Standley. 1915. Flora of New Mexico. Contrib. U.S. Nat. Mus., vol. 19, Washington, D.C.

ADDENDUM

The following interesting and relevant works appeared while our book was in the press.

Pijl, L. van der, and C. H. Dodson. 1967. Orchid Flowers; Their Pollination and Evolution. University of Miami Press, Florida.

Scheithauer, W. 1967. Hummingbirds. English transl., Th. Crowell, New York.

The Plates

PLATE I. DESERT AND SEMI-DESERT HABITATS

(A) Colorado Desert, California. *Fouquieria splendens* in foreground. (B) Same; *Belope-rone californica* in foreground. (C) Open brushland of scrub oaks on western edge of Mojave Desert. *Penstemon centranthifolius* in foreground.

PLATE 2.   ARID WOODLAND AND SCRUB HABITATS

(A) Digger pine woodland in inner South Coast Range of California. This is a typical habitat of *Penstemon centranthifolius* and the Costa hummingbird. (B) Chaparral in Coast Range of central California. *Trichostema lanatum* in foreground. (C) Sage scrub on outwash plains of San Gabriel Mountains, southern California. *Delphinium cardinale* growing intermixed with sage and other plants.

PLATE 3. COASTAL HABITATS
(A) Mountainous coastline of northern California. Typical habitat of Allen humming-bird. (B) Slopes of Santa Lucia Mountains above the ocean, central California. *Aquilegia formosa truncata* growing in the soft brushy vegetation. (C) Maritime scrub on Point Sal, central California. *Castilleja* sp. in foreground.

PLATE 4. MOUNTAIN FOREST HABITATS

(A) Open Ponderosa pine forest on plateau of central Arizona. Typical habitat of *Ipomopsis aggregata*, which is visible in foreground. (B) Hemlock-White pine forest, Sierra Nevada, California. Habitat of Calliope hummingbird and *Penstemon newberryi*. (C) Mountain hemlock and Whitebark pine at timberline on Mount Lassen, northeastern California. Habitat of Calliope hummingbird and *Castilleja payneae*.

PLATE 5. SCROPHULARIACEAE

(A) *Penstemon barbatus.*
(C) *Penstemon newberryi.*
(E) *Galvezia speciosa.*

(B) *Penstemon centranthifolius.*
(D) *Mimulus cardinalis.*
(F) *Keckia cordifolia.*

A

B

C

D

PLATE 6. CASTILLEJA (SCROPHULARIACEAE)

(A) *Castilleja payneae.*

(C) *Castilleja linariaefolia.*

(B) *Castilleja miniata.*

(D) *Castilleja chromosa.*

PLATE 7. LABIATAE
(A) *Salvia lemmoni.*
(B) *Salvia spathacea.*
(C) *Monardella macrantha.*
(D) *Trichostema lanatum.*

PLATE 8. SYMPETALOUS DICOTYLEDONS

(A) *Fouquieria splendens* (Fouquieriaceae).   (B) *Ipomopsis aggregata* (Polemoniaceae).

(C–D) *Beloperone californica* (Acanthaceae).

(E) *Anisacanthus thurberi* (Acanthaceae).   (F) *Bouvardia glaberrima* (Rubiaceae).

PLATE 9. SYMPETALOUS AND CHORIPETALOUS DICOTYLEDONS

(A) *Lonicera involucrata ledebourii* (Caprifoliaceae).

(B) *Zauschneria californica latifolia* (Onagraceae).

(C) *Ribes speciosum* (Saxifragaceae).

(D) *Astragalus coccineus* (Papilionaceae).

(E–F) *Echinocereus triglochidiatus* (Cactaceae).

PLATE 10. CHORIPETALOUS DICOTYLEDONS AND MONOCOTYLEDONS

(A) *Aquilegia formosa truncata* (Ranuncula-ceae).

(B) *Delphinium cardinale* (Ranunculaceae).

(C) *Silene laciniata* (Caryophyllaceae).

(D) *Brodiaea ida-maia* (Liliaceae).

PLATE II. COSTA HUMMINGBIRD ($\male$)
(A–B, D) Colorado Desert, Calif., March. Note hummingbird cleaning its bill on bare twig in B.
(C) Claremont, Calif., June.

PLATE 12. ANNA HUMMINGBIRD (♂)
(A–C) Claremont, Calif., February (A) and October (B, C). (D) SanGabriel Mountains, Calif., August.

PLATE 13. ALLEN HUMMINGBIRD (probably an immature ♂ of this species)
(A–D) Migrating southward through Claremont, Calif., August.

PLATE 14.   BLACK-CHINNED HUMMINGBIRD (♂ AND ♀)
(A–B) Male, Southwestern Research Station, Chiricahua Mountains, Arizona, August. (C–D) Female, same time and place.

PLATE 15.   RUFOUS AND CALLIOPE HUMMINGBIRDS

(A–B) Rufous (♂) migrating northward through Claremont, Calif., March. (C) Rufous (♂) migrating northward through high San Gabriel Mountains, Calif., during season of snows, April. (D) Calliope, Mammoth Lakes, Sierra Nevada, Calif., July.

PLATE 16.   BLUE-THROATED AND OTHER HUMMINGBIRDS
(A) Blue-throated hummingbird, Southwestern Research Station, Chiricahua Mountains, Arizona, May. (B) Rivoli hummingbird, same time and place. (C) Broad-billed hummingbird, Santa Catalina Mountains, Arizona, May. (D) Broad-tailed hummingbird, Flagstaff, Arizona, August.

PLATE 17.   BIRDS OF THE POST-BREEDING SEASON IN THE HIGH MOUNTAINS
(A–D) San Gabriel Mountains, Calif., September. The distinctive species characters are undeveloped. The bird is probably an immature Rufous or Allen.

PLATE 18. COSTA HUMMINGBIRD (♂) FEEDING ON *Penstemon centranthifolius*
(A–D) Cajon Pass, San Bernardino Mountains, Calif., May.

PLATE 19. HUMMINGBIRDS FEEDING ON *Beloperone californica*
(A) Costa (♂), Joshua Tree National Monument, Colorado Desert, Calif., March. (B) Palm Canyon, Colorado Desert, Calif., April. (C–D) Costa ?, Rancho Santa Ana Botanic Garden, Claremont, Calif., September. Note yellow Beloperone pollen on head in D.

PLATE 20. HUMMINGBIRDS FEEDING ON MOUNTAIN PENSTEMONS

(A–B) Allen (♂) on *Penstemon newberryi*, Mammoth Lakes, Sierra Nevada, Calif., July.

(C) Rufous (♂) on *Penstemon barbatus*, Rustler Park, Chiricahua Mountains, Arizona, August.

(D) Unidentified hummingbird on *Penstemon barbatus*, same time and place as C.

PLATE 21. HUMMINGBIRDS FEEDING ON *Zauschneria californica latifolia*
(A–D) Unidentified hummingbird, probably a Rufous or Allen, San Gabriel Mountains, Calif.,
September.

PLATE 22.   HUMMINGBIRDS FEEDING ON *Trichostema lanatum*

(A) Costa (♂) perched overlooking Trichostema patch, Rancho Santa Ana Botanic Garden, Claremont, Calif., May. (B–D) Birds visiting and pollinating flowers, same time and place.

PLATE 23. HUMMINGBIRDS FEEDING ON *Delphinium cardinale*
(A) Costa (♂) on perch overlooking Delphinium population, Claremont, Calif., June. (B–D) Birds visiting and pollinating flowers, same time and place.

PLATE 24. HUMMINGBIRDS FEEDING ON IPOMOPSIS AND MIMULUS
(A–B) Birds on *Ipomopsis aggregata*, Flagstaff, Arizona, August. (C–D) Anna (♂) on *Mimulus cardinalis*, San Gabriel Mountains, Calif., August. Note bird probing a flower bud in D.

PLATE 25. HEAD POLLINATION

(A) Hummingbird on *Trichostema lanatum*. Note blue pollen on bird's head. (B) Costa ♀ on *Beloperone californica*. (C–D) Anna (♂) on *Mimulus cardinalis*.

PLATE 26.   CHIN POLLINATION

(A–B) Hummingbird on *Zauschneria californica latifolia.* (C–D) Hummingbirds on *Ribes speciosum.* The bird contacts the anthers with both chin and forehead, but contacts stigma mainly with its chin. Costa (♂) in D.

PLATE 27.   CARPENTER BEES FEEDING ON HUMMINGBIRD FLOWERS

(A) Carpenter bee (Xylocopa) collecting pollen from *Mimulus cardinalis*. The bee is transferring pollen. (B) Xylocopa piercing corolla tube for nectar in *Mimulus cardinalis*. The bee does not contact the stamens and stigma in this operation. (C–D) Xylocopa piercing corolla tube of *Zauschneria californica latifolia*. The bee obtains nectar but does not contact stamens and stigma.

PLATE 28. HUMMINGBIRDS FEEDING ON BEE FLOWERS (ARCTOSTAPHYLOS)
(A) Bee (Osmia) visiting *Arctostaphylos parryana pinetorum*. (B–D) Anna hummingbirds visiting *Arctostaphylos viscida*. Some bill-tip pollination may take place.

PLATE 29. HUMMINGBIRDS FEEDING ON BEE, WASP, AND BUTTERFLY FLOWERS (PENSTEMON, LILIUM)
(A) Bee (*Anthophora californica*) on *Penstemon spectabilis*. (B) Wasp (Pseudomasaris) on *Penstemon spectabilis*. These bees and wasps are the usual pollinators of this Penstemon. (C) Anna hummingbird (♀) ? feeding on *Penstemon spectabilis*. (D) Hummingbird visiting *Lilium humboldtii*, a butterfly flower.

PLATE 30.   HUMMINGBIRDS FEEDING ON BEE FLOWERS (ISOMERIS)

(A) Flowers of *Isomeris arborea*. (B) Carpenter bee (*Xylocopa brasilianorum*) visiting *Isomeris arborea*. This and other large bees are the main pollinators of the Isomeris. (C–D) Hummingbirds feeding on *Isomeris arborea*. The hummingbird visits bring about some pollination.

Key to Equivalent Common and Latin Names of Plants and Animals Mentioned in the Text

Abeille's hummingbird : *Abeillia abeillei*
Abeillia abeillei : Abeille's hummingbird
Adenostema fasciculatum (Rosaceae) : Chamise
Agave (Agavaceae) : Century plant; Maguey
Allen hummingbird : *Selasphorus sasin*
Alpine lily : *Lilium parvum* (Liliaceae)
Amazilia rutila : Cinnamomeous hummingbird
Amazilia violiceps : Violet-crowned hummingbird
Anisacanthus thurberi (Acanthaceae) : Chuparosa; Desert honeysuckle
Anna hummingbird : *Calypte anna*
Anthophora californica (Apidae) : Anthophorid bee
Anthophorid bee : *Anthophora californica* (Apidae)
Anthracothorax prevostii : Prevost's mango hummingbird
Antirrhinum (Scrophulariaceae) : Snapdragon
Apis mellifera (Apidae) : Honey bee
Aquilegia (Ranunculaceae) : Columbine
Arbutus menziesii (Ericaceae) : Madrone
Archilochus alexandri : Black-chinned hummingbird
Archilochus colubris : Ruby-throated hummingbird
Arctostaphylos (Ericaceae) : Manzanita
Arctostaphylos glauca (Ericaceae) : Bigberry manzanita
Arctostaphylos nevadensis (Ericaceae) : Pine-mat manzanita
Arctostaphylos parryana (Ericaceae) : Parry manzanita
Arctostaphylos viscida (Ericaceae) : White leaf manzanita
Artemesia californica (Compositae) : California sagebrush
Aspen : Populus (Salicaceae)
Astragalus (Papilionaceae) : Locoweed; Rattleweed

KEY TO NAMES

Beard-tongue : Penstemon (Scrophulariaceae)
Bee-balm : Monarda (Labiatae)
Bee hummingbird : *Calypte helenae*
Beloperone californica (Acanthaceae) : Chuparosa
Bigberry manzanita : *Arctostaphylos glauca* (Ericaceae)
Bird's foot trefoil : Lotus (Papilionaceae)
Blackberry : Rubus (Rosaceae)
Black-chinned hummingbird : *Archilochus alexandri*
Black sage : *Salvia mellifera* (Labiatae)
Bladderpod : *Isomeris arborea* (Capparidaceae)
Blue-throated hummingbird : *Lampornis clemenciae*
Bombus (Apidae) : Bumblebee
Box-thorn : Lycium (Solanaceae)
Broad-billed hummingbird : *Cynanthus latirostris*
Broad-tailed hummingbird : *Selasphorus platycercus*
Brodiaea ida-maia (Liliaceae) : Fire-cracker plant
Buff-browed hermit hummingbird : *Phoethornis superciliosus*
Bumblebee : Bombus (Apidae)
Bush monkey-flower : *Diplacus aurantiacus* (Scrophulariaceae)

California buckwheat : *Eriogonum fasciculatum* (Polygonaceae)
California fuchsia : Zauschneria (Onagraceae)
California lilac : Ceanothus (Rhamnaceae)
California sagebrush : *Artemesia californica* (Compositae)
Calliope hummingbird : *Stellula calliope*
Calothorax lucifer : Lucifer hummingbird
Calypte anna : Anna hummingbird
Calypte costae : Costa hummingbird
Calypte helenae : Bee hummingbird
Cardinal-flower : *Lobelia cardinalis* (Campanulaceae)
Carpenter bee : Xylocopa (Apidae)
Carpodacus mexicanus (Fringillidae) : Linnet
Castilleja (Scrophulariaceae) : Indian paintbrush
Catchfly : Silene (Caryophyllaceae)
Ceanothus (Rhamnaceae) : California lilac
Century plant : Agave (Agavaceae)
Chamise : *Adenostema fasciculatum* (Rosaceae)
Chilicote : *Erythrina flabelliformis* (Papilionaceae)
Chilopsis linearis (Bignoniaceae) : Desert willow
Cholla : Opuntia (Cactaceae)

Chuparosa : *Anisacanthus thurberi* (Acanthaceae); *Beloperone californica* (Acanthaceae)
Cinnamomeous hummingbird : *Amazilia rutila*
Cirsium (Compositae) : Thistle
Coast lily : *Lilium maritimum* (Liliaceae)
Coerebidae : Honeycreepers
Colibri thalassinus thalassinus : Mexican violet-eared hummingbird
Columbine : Aquilegia (Ranunculaceae)
Coral-bean : *Erythrina flabelliformis* (Papilionaceae)
Coral-tree : Erythrina (Papilionaceae)
Costa hummingbird : *Calypte costae*
Crataegus (Rosaceae) : Hawthorn
Currant : Ribes (Saxifragaceae)
Cynanthus latirostris : Broad-billed hummingbird
Cyrtid fly : *Eulonchus smaragdinus* (Cyrtidae)

Delphinium (Ranunculaceae) : Larkspur
Delphinium cardinale (Ranunculaceae) : Scarlet larkspur
Desert honeysuckle : *Anisacanthus thurberi* (Acanthaceae)
Desert lavender : *Hyptis emoryi* (Labiatae)
Desert willow : *Chilopsis linearis* (Bignoniaceae)
Digger pine : *Pinus sabiniana* (Pinaceae)
Diplacus aurantiacus (Scrophulariaceae) : Bush monkey-flower
Dudleya (Crassulaceae) : Live-forever

Echinocereus (Cactaceae) : Hedgehog cactus
Elderberry : Sambucus (Caprifoliaceae)
Ensifera ensifera : Sword-billed hummingbird
Epilobium (Onagraceae) : Willow-herb
Eriogonum fasciculatum (Polygonaceae) : California buckwheat
Erythrina (Papilionaceae) : Coral-tree
Erythrina flabelliformis (Papilionaceae) : Coral-bean; Chilicote
Eugenes fulgens : Rivoli hummingbird
Eulonchus smaragdinus (Cyrtidae) : Cyrtid fly
Evening primrose : Oenothera (Onagraceae)

Figwort : Scrophularia (Scrophulariaceae)
Fire-cracker plant : *Brodiaea ida-maia* (Liliaceae)
Flannel bush : Fremontia (Sterculiaceae)
Fouquieria splendens (Fouquieriaceae) : Ocotillo
Fremontia (Sterculiaceae) : Flannel bush

Fritillaria recurva (Liliaceae) : Scarlet fritillary
Fuchsia-flowered gooseberry : *Ribes speciosum* (Saxifragaceae)

Gambel's sparrow : *Zonotrichia leucophrys* (Fringillidae)
Giant hummingbird : *Patagona gigas*
Gilia (Polomoniaceae) : Gilia
Gooseberry : Ribes (Saxifragaceae)

Hawkmoths : Sphingidae
Hawthorn : Crataegus (Rosaceae)
Hedgehog cactus : Echinocereus (Cactaceae)
Hedge-nettle : Stachys (Labiatae)
Hemlook : Tsuga (Pinaceae)
Hermit hummingbird : Phoethornis
Honey bee : *Apis mellifera* (Apidae)
Honeycreepers : Coerebidae
Honeysuckle : Lonicera (Caprifoliaceae)
Humboldt lily : *Lilium humboldtii* (Liliaceae)
Hummingbirds : Trochilidae
Hylocharis leucotis : White-eared hummingbird
Hyptis emoryi (Labiatae) : Desert lavender

Indian paintbrush : Castilleja (Scrophulariaceae)
Indian pink : *Silene californica* (Caryophyllaceae)
Indian warrior : *Pedicularis densiflora* (Scrophulariaceae)
Ipomoea (Convolvulaceae) : Morning-glory
Ipomoea coccinea (Convolvulaceae) : Star-glory
Ipomopsis aggregata (Polemoniaceae) : Scarlet gilia
Isomeris arborea (Capparidaceae) : Bladderpod

Juniper : *Juniperus californica* and related species (Cupressaceae)
Juniperus californica (Cupressaceae) : Juniper

Klais guimeti : Violet-headed hummingbird

Lampornis clemenciae : Blue-throated hummingbird
Larkspur : Delphinium (Ranunculaceae)
Leafcutting bee : Osmia (Megachilidae)
Lemonadeberry : *Rhus integrifolia* (Anacardiaceae)
Lilium (Liliaceae) : Lily
Lilium humboldtii (Liliaceae) : Humboldt lily

Lilium maritimum (Liliaceae) : Coast lily
Lilium parvum (Liliaceae) : Alpine lily
Lily : Lilium (Liliaceae)
Linnet : *Carpodacus mexicanus* (Fringillidae)
Lithocarpus (Fagaceae) : Tanbark oak
Live-forever : Dudleya (Crassulaceae)
Lobelia cardinalis (Campanulaceae) : Cardinal-flower; Scarlet lobelia
Locoweed : Astragalus (Papilionaceae)
Longuemare's hermit hummingbird : *Phoethornis longuemareus*
Lonicera (Caprifoliaceae) : Honeysuckle
Lonicera involucrata (Caprifoliaceae) : Honeysuckle; Twinberry
Lotus (Papilionaceae) : Bird's foot trefoil
Lousewort : Pedicularis (Scrophulariaceae)
Lucifer hummingbird : *Calothorax lucifer*
Lycium (Solanaceae) : Box-thorn

Madrone : *Arbutus menziesii* (Ericaceae)
Maguey : Agave (Agavaceae)
Manzanita : Arctostaphylos (Ericaceae)
Masarid wasp : Pseudomasaris (Masaridae)
Mexican violet-eared hummingbird : *Colibri thalassinus thalassinus*
Micropodidae : Swifts
Mimulus (Scrophulariaceae) : Monkey-flower
Monarda (Labiatae) : Bee-balm
Monkey-flower : Mimulus (Scrophulariaceae)
Morning-glory : Ipomoea (Convolvulaceae)
Mountain hemlook : *Tsuga mertensiana* (Pinaceae)
Mountain pride : *Penstemon newberryi* (Scrophulariaceae)

Ocotillo : *Fouquieria splendens* (Fouquieriaceae)
Oenothera (Onagraceae) : Evening primrose
Opuntia (Cactaceae) : Prickly-pear; Cholla
Osmia (Megachilidae) : Leafcutting bee
Our Lord's candle : *Yucca whipplei* (Agavaceae)

Parry manzanita : *Arctostaphylos parryana* (Ericaceae)
Passiflora vitifolia (Passifloraceae) : Scarlet Passion-flower
Patagona gigas : Giant hummingbird
Pedicularis (Scrophulariaceae) : Lousewort
Pedicularis densiflora (Scrophulariaceae) : Indian warrior

Penstemon (Scrophulariaceae) : Beard-tongue
Penstemon centranthifolius (Scrophulariaceae) : Scarlet bugler
Penstemon newberryi (Scrophulariaceae) : Mountain pride
Phacelia minor (Hydrophyllaceae) : Wild Canterbury bell
Phoethornis : Hermit hummingbird
Phoethornis longuemareus : Longuemare's hermit hummingbird
Phoethornis superciliosus : Buff-browed hermit hummingbird
Pine-mat manzanita : *Arctostaphylos nevadensis* (Ericaceae)
Pinus albicaulis (Pinaceae) : White-bark pine
Pinus monophylla (Pinaceae) : Pinyon pine
Pinus monticola (Pinaceae) : White pine
Pinus ponderosa (Pinaceae) : Ponderosa pine
Pinus sabiniana (Pinaceae) : Digger pine
Pinyon pine : *Pinus monophylla* and related species (Pinaceae)
Piranga ludoviciana (Thraupidae) : Western tanager
Pitcher-sage : *Salvia spathacea* (Labiatae)
Poison oak : *Rhus diversiloba* (Anacardiaceae)
Ponderosa pine : *Pinus ponderosa* (Pinaceae)
Populus (Salicaceae) : Aspen
Prevost's mango hummingbird : *Anthracothorax prevostii*
Prickly-pear : Opuntia (Cactaceae)
Pseudomasaris (Masaridae) : Masarid wasp

Quercus dumosa (Fagaceae) : Scrub oak

Rattleweed : Astragalus (Papilionaceae)
Redwood : *Sequoia sempervirens* (Taxodiaceae)
Rhamnus (Rhamnaceae) : Buckthorn
Rhamphomicron microrhynchum : An Andean hummingbird—we know of no common
    name for it
Rhus diversiloba (Anacardiaceae) : Poison oak
Rhus integrifolia (Anacardiaceae) : Lemonadeberry
Ribes (Saxifragaceae) : Currant; Gooseberry
Ribes speciosum (Saxifragaceae) : Fuschia-flowered gooseberry
Rivoli hummingbird : *Eugenes fulgens*
Romero : *Trichostema lanatum* (Labiatae)
Rubus (Rosaceae) : Blackberry
Ruby-throated hummingbird : *Archilochus colubris*
Rufous hummingbird : *Selasphorus rufus*

Sage : Salvia (Labiatae)
Salix (Salicaceae) : Willow
Salvia (Labiatae) : Sage
Salvia apiana (Labiatae) : White sage
Salvia mellifera (Labiatae) : Black sage
Salvia spathacea (Labiatae) : Pitcher-sage
Sambucus (Caprifoliaceae) : Elderberry
Sarcodes sanguinea (Pyrolaceae) : Snow plant
Scarlet bugler : *Penstemon centranthifolius* (Scrophulariaceae)
Scarlet fritillary : *Fritillaria recurva* (Liliaceae)
Scarlet gilia : *Ipomopsis aggregata* and related species (Polemoniaceae)
Scarlet larkspur : *Delphinium cardinale* (Ranunculaceae)
Scarlet lobelia : *Lobelia cardinalis* (Campanulaceae)
Scarlet Passion-flower : *Passiflora vitifolia* (Passifloraceae)
Scrophularia (Scrophulariaceae) : Figwort
Scrub oak : *Quercus dumosa* (Fagaceae)
Selasphorus platycercus : Broad-tailed hummingbird
Selasphorus rufus : Rufous hummingbird
Selasphorus sasin : Allen hummingbird
Sequoia sempervirens (Taxodiaceae) : Redwood
Silene (Caryophyllaceae) : Catchfly
Silene californica (Caryophyllaceae) : Indian pink
Snapdragon : Antirrhinum (Scrophulariaceae)
Snow plant : *Sarcodes sanguinea* (Pyrolaceae)
Sphingidae : Hawkmoths
Stachys (Labiatae) : Hedge-nettle
Star-glory : *Ipomoea coccinea* (Convolvulaceae)
Stellula calliope : Calliope hummingbird
Swifts : Micropodidae
Sword-billed hummingbird : *Ensifera ensifera*

Tanbark oak : Lithocarpus (Fagaceae)
Thistle : Cirsium (Compositae)
Trichostema lanatum (Labiatae) : Romero; Wooly blue-curls
Trochilidae : Hummingbirds
Tsuga (Pinaceae) : Hemlock
Tsuga mertensiana (Pinaceae) : Moutain hemlock
Twinberry : *Lonicera involucrata* (Caprifoliaceae)

Violet-crowned hummingbird : *Amazilia violiceps*

KEY TO NAMES

Violet-headed hummingbird : *Klais guimeti*

Western tanager : *Piranga ludoviciana* (Thraupidae)
White-bark pine : *Pinus albicaulis* (Pinaceae)
White-eared hummingbird : *Hylocharis leucotis*
White leaf manzanita : *Arctostaphylos viscida* (Ericaceae)
White pine : *Pinus monticola* (Pinaceae)
White sage : *Salvia apiana* (Labiatae)
Wild Canterbury bell : *Phacelia minor* (Hydrophyllaceae)
Willow : Salix (Salicaceae)
Willow-herb : Epilobium (Onagraceae)
Wooly blue-curls : *Trichostema lanatum* (Labiatae)

Xylocopa (Apidae) : Carpenter bee

Yucca whipplei (Agavaceae) : Our Lord's candle

Zauschneria (Onagraceae) : California fuchsia
Zonotrichia leucophrys (Fringillidae) : Gambel's sparrow

Index

Birds are listed under their common names, insects and plants under their Latin names. Consult the *Key to Equivalent Common and Latin Names* for cross-references.

Buff-browed hermit hummingbird, 93

Calliope hummingbird, 4, 10-13, 15-16, 26, 32, 51-53, 67-68, Plates *4, 15*
Castilleja: *C. affinis*, 19; *C. angustifolia*, 19, 43; *C. applegatei*, 19; *C. austromontana*, 19, 43; *C. brevilobata*, 19; *C. breweri*, 19, 51-52, 72; *C. chromosa*, 19, 48, Plate *6*; *C. covilleana*, 19; *C. crista-galli*, 19; *C. cruenta*, 19, 43; *C. culbertsonii*, 19; *C. elmeri*, 19; *C. exilis*, 19, 56; *C. foliolosa*, 19, 48; *C. franciscana*, 19; *C. fraterna*, 19; *C. haydeni*, 19; *C. hispida*, 20; *C. hololeuca*, 20; *C. inconstans*, 20; *C. integra*, 20; *C. lanata*, 20, 43; *C. latifolia*, 20; *C. laxa*, 20, 43; *C. lemmonii*, 20; *C. leschkeana*, 20; *C. linariaefolia*, 20, 51, Plate *6*; *C. martinii*, 20, 48; *C. miniata*, 20, 51-53, 72, Plate *6*; *C. minor*, 20, 56; *C. nana*, 20; *C. neglecta*, 20; *C. organorum*, 20; *C. parviflora*, 20; *C. patriotica*, 20, 43, 53; *C. payneae*, 20, 51, Plates *4, 6*; *C. peirsonii*, 20; *C. plagiotoma*, 20; *C. pruinosa*, 20; *C. rhexifolia*, 20, 43; *C. roseana*, 20; *C. rupicola*, 20, 43; *C. stenantha*, 20, 33, 56; *C. suksdorfii*, 20, 43; *C. subinclusa*, 20; *C. uliginosa*, 20; *C. wightii*, 20; *C. wootoni*, 20
Centaurium venustum, 37
Chaparral habitat, 48, Plate *2*
Chilopsis linearis, 47
Chin pollination, 30-31, Plate *26*
Cinnamomeous hummingbird, 9
Co-adaptation, 1-3, 86ff.
Coastal sage scrub habitat, 48, Plate *2*
Coast Ranges of California, 48-50, Plates *2, 3*
Coerebidae, 89
Collomia rawsoniana, 18

Colorado Desert habitat, 45-47, Plate *1*
Common red coloration: explanation of, 81ff.; prevalence in western American hummingbird flowers, 77ff.
Competition, 52, 74, 87, 90, 93-94
Cronquist, A., 17
Cross-pollination: in *Delphinium cardinale*, 35ff.
Costa hummingbird, 4, 10, 12-13, 15, 36ff., 47-48, 50, 68, Plates *2, 11, 18, 19, 22, 23, 25, 26*

Darwin, Ch., v
Davis, J. S., 58
Davis, R. J., 17, 40
Dawson, W. L., 13, 49
Delphinium: *D. cardinale*, 23, 30, 33, 35ff., 48, 59, 71, Plates *2, 10, 23*; *D. nudicaule*, 23, 33, 43, 49, 59
Delpino, F., 77
Diplacus: *D. aurantiacus*, 20, 43, 49, 50; *D. flemingii*, 20; *D. longiflorus*, 48; *D. puniceus*, 20, 33, 48, 69
Dudleya: *D. cymosa minor*, 64, 89; *D. lanceolata*, 37, 89

Echinocereus triglochidiatus, 23, Plate *9*
Epling, C., 35
Eriastrum sapphirinum, 37
Eriogonum fasciculatum, 48
Erythrina flabelliformis, 23, 43, 61
Eulonchus smaragdinus, 48
Exclusiveness, development and spread of, 89ff.

Faegri, K., v, 58
Floral mechanisms: attractive, 25-27, 88-89; exclusive, 25, 27, 62, 89ff.; pollination, 4, 25, 30-32, 34ff., 88-89; protective, 25, 28-30